January 1982.

With love to a good
and Faithful Friend.

Maureen FCJ

WILLIAM WHISTON

WILLIAM WHISTON

Maureen Farrell

ARNO PRESS

A New York Times Company
New York • 1981

Editorial Supervision: RITA LAWN

First Publication 1981 by Arno Press Inc.
Copyright © 1980 by Maureen Farrell
Reproduced by permission of Maureen Farrell

THE DEVELOPMENT OF SCIENCE: Sources for the History of Science
ISBN for complete set: 0-405-13850-4
See last pages of this volume for titles.

Manufactured in the United States of America

Library of Congress Cataloging in Publication Data

Farrell, Maureen.
  William Whiston.

  (The Development of science)
  Originally presented as the author's thesis,
University of Manchester, 1973, under the title:
The life and work of William Whiston.
  Bibliography: p.
  Includes index.
  1. Whiston, William, 1667-1752. 2. Church of
England--Clergy--Biography. 3. Baptists--Clergy--
Biography. 4. Clergy--England--Biography. I. Se-
ries: Development of science.
BX5199.W52F37 1981   509'.2'4 [B]   80-2088
ISBN 0-405-13854-7

THE LIFE AND WORK OF

WILLIAM WHISTON

Will:ᵐ Whiston
A. M.

THE LIFE AND WORK OF

WILLIAM WHISTON

by

Maureen Farrell F.C.J.

Thesis submitted to the Faculty of Technology
of the University of Manchester
for the degree of

DOCTOR OF PHILOSOPHY

April 1973.

During the years 1970-73 when I was engaged in full-
time research centred on William Whiston, Professor I. Bernard
Cohen was a constant source of encouragement. It is with great
pleasure that I now respond to Professor Cohen's invitation to
accept Arno Press as publisher for this thesis. My external
examiner, Eric Forbes, professor of history of science at
Edinburgh University, has suggested that at some future date
the material could be re-written in chronological form providing the biography of a colourful figure of the eighteenth
century for a wide readership.

Since the completion of this thesis I have had the opportunity
to present seminars on Whiston to varied audiences in scholarly
circles and to amateur astronomical societies and local history
groups. The tercentenary of the Greenwich Observatory in 1975
offered another opportunity to highlight Whiston's contribution
to the longitude problem. The paper read on this occasion appears
in Vistas in Astronomy, Volume 20, 1976, Pergamon Press. On the
same occasion a magnificent exhibition was mounted in the Queen's
House in Greenwich Park and under the direction of Derek Howse I
prepared a section relating to Whiston's activities. [First West
Room (1675-1720), Section 14. Handbook on the Exhibition, edited
by Colin A. Ronan].

Printed charts of astronomical interest designed by Whiston
are now scattered in many distant archives. The unique collection
existing in the Houghton Library, Harvard, prompted me to write
the paper, Rare Items Relating to William Whiston (1667-1752) in
the Houghton Library, Harvard Library Bulletin, July 1976.

During the period January-December 1981, I look forward to
a period of study leave from UMIST and I hope to use this time

to further my study of Whiston and his contribution to the eighteenth century scene. I owe a debt of gratitude to Professor Donald Cardwell not only for his constant encouragement but also from time to time for his mild reproaches urging me to devote further time and energy to a fuller treatment of Whiston's achievements and to a study of his interaction with contemporary men of learning.

It is my hope that the publication of this work will help to heighten interest in this remarkable man. At the Greenwich celebrations in 1975, I called my paper The Longitude Man. Church historians would recognise him as the Josephus man, who translated the works of the Jewish historian, Flavius Josephus, from the original Greek into English. Whiston's enthusiastic attempts to found a society for the practice of Primitive Christianity resulted from his involvement in the Trinitarian debate of the early eighteenth century. Cosmologists recall his New Theory of the Earth, 1696, as a fascinating sequel to Thomas Burnet's Sacred Theory of the Earth, 1681-84. Oliver Goldsmith has immortalised Whiston's memory in The Vicar of Wakefield, a tale of successively devastating miseries overtaking the ever hopeful clergyman. I trust that this present work will reveal Whiston as a man of considerable intellectual stature, a person of enthusiasm and with wide-ranging interests, one whose charm over-rode his eccentricities and who considered his own greatest claim to fame to be his tenure of the Lucasian Professorship of Mathematics in Cambridge in succession to Isaac Newton.

Maureen Farrell, FCJ,
Department of History of Science & Technology,
UMIST, Manchester.

3rd July, 1980

ACKNOWLEDGMENTS

The preparation of this Thesis has brought me into contact with many persons who have gladly shared their knowledge and helped me in many ways.   The staff in libraries, archives and record offices and members of the clergy in Lowestoft, Drayton (Norfolk), Norton-juxta-Twycross and Lyndon (Rutland) have helped me to locate the material for this Thesis.   John Gilmour and Charles Parkin at Cambridge have given me great encouragement over several years and I was privileged to have access to the magnificent collection of Whiston's works at Clare College, Cambridge. Through the kindness of Mr. and Mrs. J.M. Conant, I visited Lyndon Hall, Rutland, where Whiston spent many happy years and where his portrait occupies a place of honour today.

My special thanks are due to the members of the Department of The History of Science and Technology at U.M.I.S.T. who encouraged me in the effort of making an ordered study of the great maze of Whiston's printed works. In particular I would like to express my appreciation of the supervision and guidance of Dr. D.S.L. Cardwell and Dr. W.V. Farrar and the services of Mrs. Emily Cooper, Departmental secretary.

Finally, I would like to thank my superior, Mother Raphael Conran for giving me the opportunity to undertake this research and my own Community at Sedgley Park College for their continued interest and support.

ABSTRACT

William Whiston was born at Norton-juxta-Twycross,
Leicestershire in 1667.   He studied first at home and
later at Tamworth Grammar School and in 1686 was admitted
to Clare Hall, Cambridge where he qualified as B.A. (1690),
and M.A. (1693), and was elected Fellow in 1691.   William
Lloyd ordained Whiston at Lichfield in 1693 and he married
Ruth Antrobus in 1699.   Whiston's New Theory of the Earth
(1696) attracted considerable attention, where he postulated
the origin of the earth from the atmosphere of a comet and
all major changes in the earth's history were attributed to
the action of comets.   Successively chaplain to Bishop
John Moore at Norwich and rector of Lowestoft, Whiston
succeeded Newton as Lucasian Professor of Mathematics
1702-1710.   The printed version of Whiston's astronomical
and mathematical lectures comprised a commentary on the
Principia and were widely used as textbooks during the
eighteenth century.   He was deprived of his Chair at
Cambridge because of his heretical tenets and went to
London where he continued his activities in religious and
scientific fields.   In 1714, Whiston was instrumental in
the establishment of the Board of Longitude and for the next
forty years made persevering efforts to solve the longitude
problem.   He gave courses of demonstration lectures on
astronomical and physical phenomena and engaged in many
religious controversies.   Much of his time was devoted to
a revival of Primitive Christianity based on the Apostolical

<u>Constitutions</u>.   Whiston mistakenly believed this work to
be an authentic document of the early Church.   It was in
fact a compilation from the period 340-380.   Whiston's
remarkable output of printed works spans the last fifty
years of his life.   Although one must concede areas of
stubbornness in his religious beliefs, the total picture is
of a man of integrity, ceaseless activity and considerable
achievement.   He had a happy family life and died in
Lyndon Hall, Rutland, in 1752, at the home of his son-in-
law, Samuel Barker.

ACADEMIC CAREER

I am a member of the religious order devoted to educational works known as the Faithful Companions of Jesus. The years 1954-1958 were spent at University College Dublin, where I was awarded B.Sc. (General) with distinction, and the Higher Diploma in Education (First class honours). During the period 1959-1965, I was a member of the Science Department at Sedgley Park College of Education, Prestwich, and undertook teaching duties in General Science and method courses. A year's research work in the Chemistry Department of Manchester University was spent investigating the polymerisation of cyclic ethers using high vacuum technique and organo-metallic catalysts. This work led to the degree M.Sc. in December 1966. The four years 1966-1970 were spent as head of the Science Department at Sedgley Park College. In 1965 I was awarded the Westminster Diploma in Sacred Studies. For the last three years I have been a full-time research assistant in the Department of History of Science and Technology, U.M.I.S.T. This Thesis is the result of these three years' research.

No portion of this Thesis has been previously submitted in support of an application for any degree of any University.

CONTENTS   <u>Page</u>

LIST OF PLATES

C H A P T E R   O N E

BIOGRAPHICAL NOTES

C H A P T E R   O N E

BIOGRAPHICAL   NOTES

1.1   BACKGROUND   TO   WHISTON'S   EARLY   YEARS

The year 1667 was not a peaceful time for the
Cavalier Parliament (1661-79).   In June 1667 Dutch ships
sailed up the Thames and destroyed English ships in the
Medway;   England was forced to make peace with the Dutch.
Edward Hyde, Earl of Clarendon, who had served Charles II
so loyally both before and after the restoration of the
monarchy in 1660 was accused of treason by Parliament and
went into exile in 1667.   Charles II took over the direction of the Government, helped by five advisers, Clifford,
Arlington, Buckingham, Ashley and Lauderdale, known as the
CABAL Administration.

The toleration of worship which Charles had promised
in the Declaration of Breda did not materialise.   At first
it looked as though the Anglicans and Presbyterians would
be reconciled.   The Presbyterian leaders were prepared to
accept a modified form of episcopacy on the lines traced
out by James Ussher, Archbishop of Armagh, in which the
bishops were to have synods associated with them.   Charles
issued a declaration in favour of this arrangement in
November 1660 and high preferments were offered to some of

NOTES: Dates prior to 1752 have been amended so as to
count the beginning of the year on 1st January.

Spelling has in many places been modernised to
assist the reader and to render the text
intelligible.

the chief Presbyterians.   The Savoy Conference met at the
Bishop of London's lodgings to work out the forms of
worship and belief.   Richard Baxter, the foremost
Presbyterian at this Conference, had great qualities of
faith and leadership but lacked powers of negotiation.
Before the end of the Savoy Conference, the Cavalier
parliament met and its decisions with regard to religious
matters were characterised by a narrow exclusiveness.
The ideal of comprehension was abandoned and the jurisdiction of ecclesiastical courts was revived.   The Corporation
Act (1661) excluded from municipal corporations all those
who refused to take the sacrament according to the rites
of the Church of England.   The Act of Uniformity (1662)
authorised once more the Prayer Book and the various
formularies of belief.   It restricted the holding of
benefices to episcopally ordained clergy.   The Conventicle
Act (1664) imposed penalties for attendance at worship not
in Anglican forms.   In 1665, the Five Mile Act forbade
any non-conformist minister to live or visit within five
miles of any corporate town or any place where he had acted
as minister.[1]

All these Acts were directed against the non-Conform-
ists possibly because their meetings were considered to be
the breeding grounds of new rebellions.   The result of
all this legislation was to emphasise religious divisions
in England.   The established Church was anxious to purge
itself of heretics and fanatics by oaths and subscriptions.
On St. Bartholomew's Day 1662, 'The great ejection' took
place, and nearly one thousand incumbents who were unwilling

to accept the liturgy or doctrines of the Church of England vacated their livings.*

The English Presbyterians, who had been the nearest rivals of the Church of England were hardest hit by the legislation against dissenters. During the eighteenth century, many of their remnant became unitarians.

In 1659, when the Royalist cause still faced an uncertain future, Edward Hyde tried desperately to secure the Church against its own immediate peril, namely the lapse of the episcopal order. Hyde despatched urgent letters to Barwick and brought pressure to bear on the exiled King to force the Bishops to action. Anglican agents were employed in what may be described as an offensive against the dominant Presbyterian party.

The plan of offering bishoprics and deaneries to leading Presbyterians was a weapon of attack on Presbyterian solidarity to be used entirely in Anglican interests.[3]

In the first few months following Charles' return to England, many petitions were addressed to the King begging him to accommodate Presbyterians in the new church settlement. Josiah Whiston, father of William Whiston, is recorded as a member of the clergy in Leicestershire who addressed Parliament in 1659 and expressed satisfaction at the restoration of the clergy to their livings.

---

* A careful perusal of the evidence leads to the conclusion that 936 ministers were ejected in 1662 in addition to 760 already ejected in 1660. Of these, at least 420 were in full Anglican orders before the Civil War and 45 more after the Restoration. (2)

> "They did acknowledge it, as the product of
> Divine Love and Goodness towards the Nation;
> that notwithstanding the many changes of
> Persons and Government, a Godly and Preaching
> Ministry, had been, and still was countenanced,
> protected and maintained by the Parliament:
> and that the Lord had been pleased after so
> many years interruption, to restore them again
> to their places, for the accomplishing of all
> those just and good things, which they formerly
> prosecuted, in order to a happy settlement."   (4)

Later in his life, Josiah Whiston was to express regret at having signed this petition, and writing to Richard Baxter he also expressed sorrow at the Civil War which preceded the execution of Charles I.

> "... the Parliament and all that assisted them
> or approved of what they did in that war against
> the King were guilty of those great sins of
> sedition and rebellion and of all the blood which
> has been shed and all the mischiefs that have
> followed and that all those troubles which have
> befallen the professing party are the expressions
> of the wrath of God chiefly for those sins which
> will never be removed till the sinners have
> acknowledged and repented of."   (5)

William Whiston was born in the village of Norton-juxta-Twycross in Leicestershire on 9th December 1667. He was the second son of the six sons and one daughter who survived to adulthood.*   Owing to his father's blindness, William acted as his amanuensis and must have spent much time in his company.

Even in July 1670, when his son William was not yet three years old, Josiah complains about his poor health and failing eyesight.

---

* The church at Norton-juxta-Twycross is listed as an
  historical monument and will always be preserved.
  The rectory is now the administrative headquarters of
  Twycross Zoo Park.

> "I have continually looked death in the face for
> these last three years, having been troubled
> with a distemper in my throat, such a narrowness
> that I have been in danger of suffocation every
> mouth I have eaten."

> "I know not whether you will be able to read these
> lines my eye-sight hath always been bad but of
> late years it waxed so dim that I have much ado,
> either to write or read." (6)

Josiah Whiston must have been inwardly troubled whether to align his political and religious loyalties with his brother Joseph who was a convinced dissenter and chaplain to Colonel Thomas Harrison the regicide[7] or to conform to the royalist tendencies within his father-in-law's household where he then resided.

Josiah Whiston used to keep the anniversary of the death of Charles I, January 30th, as a day of solemn fast. He made provision in his Will of 19th February 1676 for his children, and left William his library.[8] At this time, Josiah's eyesight must have been very poor as he did not sign the Will, but made a mark in the presence of a witness. On 30th December 1685 he revised his Will and expressed his desire that William should enter the ministry of the Church.

> "Secondly, whereas my foresaid last Will and
> Testament I have given unto my son William only
> the sum of Fifty pounds, I do now by these
> presents give and bequeath unto my said son
> William the sum of thirty pounds more for an
> addition unto his portion and do beg of God
> Almighty and of other good friends that a way
> may be found out for his education at the
> University in order to his doing God service in
> the work of the ministry to which I have as well
> as I am able dedicated him, beseeching Almighty
> God to give him grace and ability that he may be
> an able minister of the New Testament." (9)

In this final version of his Will, Josiah had to reduce the sum of money left to each of his children with the exception of William.   One wonders what misdemeanours had been committed by Josiah's eldest son, Rosse, whose bequest was reduced from twenty pounds to five shillings,

> "because he hath an estate left to him by his grandmother Mrs. Katharine Rosse, and for some other reasons best known to myself."   (10)

The total of all his goods and chattels, including crops in cultivation amounted to £1,093. 14s. 0d., but about £650 was in the hands of debtors and this may have accounted for the reduced circumstances in which the family found themselves.(11)

William refers to himself as being of a weak constitution and subject to attacks of melancholy depression. Perhaps because of this he did not leave home to go to Queen Elizabeth Grammar School at Tamworth until he was seventeen years old.

There is evidence of a school at Tamworth since 960 A.D. during the reign of King Edgar.   It was refounded by Elizabeth in 1588.(12)   There are no extant records of the curriculum followed in Whiston's day but there was a sound tradition of Latin teaching.   Samuel Shaw, headmaster before George Antrobus, wrote an Epitome of the Latin Grammar and Simm's Bibliotheca Staffordiensis also ascribes to Samuel Shaw a Latin Syntax for Tamworth School.(13)   The Reverend George Antrobus raised the school to a great pitch of prosperity.(14)   During his headmastership the academic side of the school was in capable hands.   Extensions were

made to the school buildings which were duly opened in
1678.   His efforts to enhance the standing of the school
must have been appreciated as numerous bequests were made
to the school in his time.   Whiston was a pupil at
Tamworth Grammar School from the autumn of 1684 to summer
1686.

Whiston's name is daily to be heard in the school as
he is jointly commemorated in the title of one of the
houses of the school, Ferrer's - Whiston's.[15]

In 1699, William married Ruth, one of the daughters
of George Antrobus, headmaster of Tamworth School.   It is
unlikely that their association dates from his school days
at Tamworth as the heavy programme of classical studies
would have left William little time or opportunity for
searching out his prospective bride.

Josiah Whiston died during William's time at Tamworth.
It is not clear what the fortunes of the Whiston family
were after 1685.   They perhaps remained in the rectory
at Norton and appointed a Vicar to do the work of the parish.
A note written by Whiston in 1696 is headed "Tamworth".[16]
When Whiston made his Will in 1734 he referred to his farm
at Dullingham Lay,[17] Cambridgeshire.   It may be that
the family moved after 1685 from Leicestershire to
Cambridgeshire.

On 30th June 1686, William was admitted to Clare Hall
as a sizar though he did not come into residence until the
September following.*

---

* Daniel Whiston (brother to William) was later a student
  at Clare Hall.   Two of Whiston's sons, William and
  George were also students there.

1.2   LIFE AT CLARE HALL

Whiston shared a room with John Laurence (1668-1732) later a clergyman, and noted for his book on gardening, The Clergyman's Recreation (1714).[18]   Whiston and Laurence remained friends for life.   The period at Clare Hall under the mastership of Samuel Blithe (Master, 1678-1713) must have been a formative period in the lives of these two young men.   Laurence was one of those churchmen who looked to Samuel Clarke as their leader and his friendship with Whiston brought him into contact with the controversial issues of the time, such as the anti-Trinitarian debate;   both Whiston and Clarke were examined at different times before the Upper House of Convocation.   John Laurence is thought to be the author of the work An Apology for Dr. Clarke containing an Account of the late Proceedings in Convocation upon his writings concerning the Trinity, London (1714).[19]   Laurence dedicated his work, A New System of Agriculture (1726) to the Princess of Wales.*

Admission to a Cambridge College as a sizar reduced the cost of University education for the poorer students.**

---

*   Queen Caroline (1682-1737) was wife of George II and daughter of John Frederick, Marquis of Brandenburg Ansbach and of Eleanor Louisa, his second wife.   She married in 1705 and was crowned Queen Consort in 1727. She came to England in 1714.   She met Leibnitz in Germany and through Samuel Clarke carried on a correspondence with Leibnitz about Newtonian philosophy. Whiston used to visit Queen Caroline at Windsor where she gathered a circle of men of letters.   (20) Laurence and Whiston may have gone together to such discussions.

**   A sizar was nominally a servant to the wealthier students but the duties of a sizar varied widely.

The status of sizar was not necessarily the same for all Colleges of the University at this period. Whiston was sizar to Nathaniel Vincent. Vincent was known to be a great friend of King James II and in 1688 this friendship put him in danger of being attacked by the Cambridge mob who were in favour of installing William of Orange on the English throne.[21] During the period 1658 to 1713, thirty seven out of ninety one students elected to fellowships at Clare Hall were admitted as sizars. The majority of Clare Fellows whose names are remembered today were admitted as sizars: Samuel Blithe, John Moore (Bishop of Norwich and Ely), Richard Laughton, William Whiston, Robert Green and Thomas Seaton.[22]

Whiston's portrait by Sarah Hoadley surveys the scene in the Combination Room at Clare today. In 1693 the new North range, comprising the Hall, Buttery, Kitchen, Combination Room and Library was completed, and Whiston would have witnessed the completion of this work begun in 1683. Edmund Prideaux's drawing[23] gives a fair idea of the appearance of the College buildings but the completion of the court and the building of the Master's Lodge was not finished until 1715.

Whiston says of the state of the College buildings,

"I was admitted to Clare-Hall, Cambridge, about the middle of 1686, while a very small part of the old College was standing: Though I question whether any of it was standing when I came to reside, which was the September following." (24)

A student sought admission through a Fellow. In the year 1658, the majority of Fellows had pupils. By 1678

when Samuel Blithe became Master, he had a virtual monopoly
of all pupils.   When the Master ceased to take new pupils,
Robert Herne became for a time practically the only tutor
and Whiston was one of his pupils.[25]   A candidate for
admission was examined as to his fitness and an admission
fee was paid.   The academic dress was the undergraduate
gown.   There were great variations in materials and costs.
The gowns were buttoned at the neck and were sometimes
lavishly adorned with up to ten dozen buttons and loops.[26]
It was common for caps to be worn and square caps seem to
have come into fashion between 1669 and 1679.   The accounts
for Whiston's undergraduate days do not survive but he tells
us himself,

> "that though I was a Pensioner for the last half
> year, yet did my whole expences for the three
> years and half till my first Degree inclusive
> not amount to so much as 100 pounds."

This in itself was no inconsiderable sum when it is to be
remembered that a clergyman of moderate means could expect
an income of only about £60 annually.*

In his Common place book[27] Whiston gives several
lists of books in his possession, the most interesting
being 156 books which he had in his room at Cambridge,
February 1688.   (See Appendix II to this chapter).

During Whiston's time at Cambridge, the Cartesian
system of natural philosophy was commonly taught in the
university.

---

* The recommended work for a discussion of student life
at Clare Hall is, W.J. Harrison: <u>Life in Clare Hall</u>,
<u>Cambridge (1658-1713)</u> (1958), Heffer, Cambridge.

"After I had taken Holy Orders (1693), I returned
to the College, and went on with my studies
there, particularly the Mathematics and the
Cartesian philosophy which was alone in vogue
with us at that time. (28)   In 1697, Samuel
Clarke published a translation and notes upon
Rohault's *Physics*.   Rohault's text was based
entirely on Cartesian principles but Samuel
Clarke introduced such notes into his translation as to lead students insensibly and by
degrees to an awareness of the philosophical
principles contained in Isaac Newton's
Principia."   (29)

The main items of expenditure were food, clothing
and books.   Whiston may have been economical in food and
clothing but he had a valuable and extensive collection of
books, most of which were purchased during his first two
years at Clare Hall.

Students were required by Statute to attend public
lectures with regularity.   In course of time attendance
ceased to be enforced, possibly because there were no
lectures.   It was then required that at least half of
each term should be kept by residence.   The original
requirement that twelve terms should be kept for a complete
B.A. degree was later modified to ten.   It would seem that
academic tests degenerated into mere formalities except for
the more competent students.[30]

The public lectures (*in scholis*) were held under the
auspices of the University in the old Schools, a series of
buildings near the present Senate House.   Private lectures
were those held by the College in the dining hall, the
Chapel or in a Tutor's rooms.   There were by statute four
public lectures a week in theology, civil law, medicine and
mathematics, while lectures in language, philosophy,
dialectic and rhetoric were five in number.[31]   The

lectures lasted from seven to eight o'clock in the morning.

John Cowell, Master of Trinity Hall, in his Will of
1611 provided for the perpetual maintenance of a logic
lecture in his College.   The lecture was to be read four
days a week for two hours every day.   The first hour or
'repetitio' was an examination of the former day's lecture.
The second or 'praelectio' a new lecture given in such a
deliberate manner that the students could take verbatim
notes.[32]

Whiston refers to,

> "the accustomed Exercise in the Schools for my
>   first degree,"   [33]

when he defended Dr. Burnet's <u>Sacred Theory of the Earth</u>.
This was probably a disputation fulfilling the University
Statute that the student had to appear four times in the
schools during his four years as an undergraduate, twice
as answerer or defendant and twice as objector.[34]

Evidence for the existence of this form of disputation
occurs in Dr. Benjamin Hoadley's <u>Account of the Life</u>,
<u>Writings and Character of Dr. Samuel Clarke</u> which is pre-
fixed to the 1731 Folio edition of Clarke's works.   In
1709, Clarke defended his doctoral thesis which was an
elaborate discourse on the question "No article of Christian
faith, delivered in Holy Scripture is disagreeable to right
reason."   Dr. James, then Regius Professor of Divinity,
a very learned and acute disputant, who,

> "After having sifted every part of Dr. Clarke's
> thesis with the strictest nicety, pressed him
> with all the force of syllogism in its various
> forms."
> "In the course of the syllogistic disputation,

he replied readily to the greatest difficulties
proposed; and pressed him so close and hard with
clear and intelligible answers; that perhaps
never was such a conflict heard in the schools."[35]

Whiston was present on this occasion and recalled the
incident clearly.[36]

The fee payable to the College for the degree of
Bachelor of Arts or Law was £3. 3. 4. In Whiston's time
at Clare Hall, a fee of ten shillings was payable by a
B.A. to the Fellow who, as Father, presented him for his
degree.[37] Whiston received the B.A. degree in the early
months of 1690, probably in January of that year. The fee
for an M.A. degree was £5. 6. 8. and Whiston found that
Bishop John Moore's gift of £5 in 1693 when he received his
degree was a considerable help in raising the required
sum.[38]

Whiston was elected to the Exeter fellowship on July
16th 1691.[39] A meeting of Fellows was held in the Ante-
Chapel for the purpose of making nominations to vacant
fellowships. The Master could veto a candidate favoured
by the Fellows and the Fellows could veto one favoured by
the Master. Three Exeter Fellowships were founded by the
Earl of Exeter in 1612 by the gift of a rent-charge of £108
a year. Each Fellow received a weekly allowance of 5s. 9d.
during residence and the Fellow was required to be ordained
after having been Master of Arts for four years.[40]

In February 1693, Whiston was promoted to Probationary
Senior Fellow of the College.[41] Probationary Fellows,
of whom there were never more than three at any one time,
enjoyed free rooms and free Commons. These Fellows

succeeded to Clare Fellowships as vacancies occurred.[42]
The average age of the Fellows in 1700 was twenty-five.
There were in Whiston's time a total of 18 Fellows and 3
Probationary Fellows with a total College population not
exceeding a hundred students.

In the Fellows Accounts kept by Samuel Blithe as
Master of Clare Hall (1678-1713), Whiston's entries
survive.[43]   They cover the period July 16th 1691 when
Whiston was elected to the Exeter Fellowship until June 8th
1699 when his Fellowship expired on his marriage to Ruth
Antrobus.   A careful study of these closely inscribed
columns of parchment reveals details about Whiston's movements during these years.   Whiston acted as one of the
Library Keepers and, for this received a half yearly salary
of £2. 10. 0.   In the autumn of 1694, Whiston moved into
the Priests' Chamber at the College, the third under Mr.
Troughton's Chamber.   In January 1695, Whiston owed
£42.19.9¼ to the Master.   The following extract shows the
various sources of income which Whiston had at this period
and after January 1694, the accounts were always concluded
in Whiston's favour.

Mr. Whiston's debet for the quarter at Christmas '93			12.00.00
add your buttery bill at Lady day	1694		18.04.11¼
add your buttery bill at Midsummer	1694		33.03.04
Income into the Priests' Chamber			3.00.00
add your buttery bill at Midsummer	1694		33.16.01
Sum totall is			100.04. 4¼
Wages & Commemoration at Lady day 1694			00.07.04
Wages & Commemoration at Midsummer 1694			00.07.00
Corne dividend at Midsummer to yourself			21.08.09

Improved rents for Ely, Littleport and Downe till Mich. 1698	00.12.08¼
Wages & Livery money at Mich. 1694	01.08.04
Deans wages for 3 quarters at Mich. 1694	01.10.00
Your sizar Martyn's Corne Rent at Midsummer 1694	02.05.00
Fines and seals for Littleington Parsonage, Rose Garden Close, Robinson's Tenants in Ely, Joyces tenants in Charlston, Warwicks Tenant in Ely, Littleport Parsonage & Lollworth for me	17.15.03¼
November 16th 1694, Mr. Laughton then paid to Geo. Labitt ten pounds of Mr. Whiston's money which I do allow him (it being paid by my orders)	10.00.00
Wages & Commemoration at Christmas 1694	00.07.04
Surplus in the buttery at Christmas 1694	01.02.10¼
January 21st 1694(1695) Recd of Mr.Laughton fourty two pounds, nineteen shillings & nine pence farthing	42.19. 9¼

On page 147 of Blithe's account book, a note written
by Whiston to the Master is pinned which reads:-

Tamworth   October 26 1691(?)

"I do hereby desire and empower ye Reverend Dr.
Blyth ye Master of our College to pay all that
is or shall become due to me as Fellow, during
my absence from the College, to Mr. Rich:
Laughton Fellow; whose discharge for the same
shall be as sufficient as if my self had given
it at any time.

Witness my hand

Will. Whiston.

To the Reverend Sam: Blyth D.D. Master of
Clare Hall Cambridge"

Richard Laughton was the best known tutor of the
College at this time and Whiston is proud to own him as his
'bosom friend'.[44]

In 1693, Whiston was conferred as Master of Arts and
then set himself up as a tutor at Clare to take pupils.
The account book at Christmas 1693 charges him £13.14.07¼ as

the buttery bill for himself and his eleven pupils.   One of
Whiston's pupils was the nephew of John Tillotson, Arch-
bishop of Canterbury, who himself had been educated at
Clare Hall.*

Whiston's ten years of residence at Clare Hall were
a formative influence in his life.   It was during these
years that he first publicly appeared in a controversial
situation namely the election of Richard Laughton to a
Clare Fellowship in preference to John Blithe, nephew of
Samuel Blithe, Master of Clare Hall.[46]

Whiston includes in his memoirs a paper entitled
Emendanda in Academia in which he suggests various reforms
relating to elections to Fellowships and College discipline.
One of these regulations is that

> "None in Holy Orders, nor Under-graduates to go
> to taverns or publick houses at all, without
> particular business with strangers there, and
> at early hours.   Others to be restrained from
> much frequenting the same."   (47)

Doubtless there was behaviour among students which
called for reform, but Whiston in his melancholy and
ponderous style takes delight in calling down dreadful
punishments on those who sought excessive pleasure.   On
July 15th 1691, on the occasion of Mr. Hollis's death, and
Whiston being chosen Fellow the next day, he wrote:

> "Affrighting is the approach of death in any
> shape, but when it comes on a sudden, in the
> midst of jollity and drinking, debauchery and

---

* Tillotson was a confidential friend of William and Mary
  and was given the see of Canterbury at the deposition of
  Sancroft in 1690.   Tillotson was strongly disliked by
  the non-jurors who refused to swear allegiance to
  William and Mary believing this to be incompatible with
  their oath to James II.   (45)

merry company, and on a sudden seizes the
trembling sinner, and in a very little time
hurries him of this world, how much more
terrible must it be .... This sudden
Providence and surprizing accident of Mr.
Hollis's Death (who was merry enough but a week
before at the commencement) seems providentially
disposed for my warning and caution, just upon
my advancement to a fellowship; not to be proud
and conceited, forgetful of God, and unmindful
of Eternity." (48)

No one could accuse Whiston of excessive jollity,
drinking and debauchery but whether he heeded his own warning "not to be proud and conceited" is not so certain.

1.3 FELLOW OF CLARE. CHAPLAIN TO BISHOP JOHN MOORE

A year after Whiston arrived in Clare Hall, he records the death of Dr. Henry More of Christs College.(49)
He was the last surviving member of the group of Cambridge
scholars collectively known as the Cambridge Platonists who
saw the use of reason and the exercise of virtue as the
twin spheres in which we enjoy God. They asserted that
faith in revelation was not incompatible with confidence
in reason and they adopted a mediating position between the
Laudians and the Calvinists. They were receptive to knowledge in all forms, but it had to prove its power to liber-
ate men's minds and enrich his spirit. In the Cambridge
Platonists there was a strain of mysticism which preserved
them from narrow rationalism. They rose to an apprehension
of God in and through nature, not beyond it. Theirs was an
attempt to assert the Christian faith in a form at once
intellectually defensible and spiritually satisfying, able
to protect religion against superstition and atheism.(50)

Henry More was one of those who initiated a tradition which powerfully affected English thought for the next two centuries, and must surely have affected Whiston during his first year at Clare Hall.*   In the year of More's death, Whiston wrote one of his meditations entitled <u>On the Reasonableness of Religion</u>[51] which echoes a primary tenet of the Cambridge Platonists.

Whiston makes frequent references in his writings to sermons and opinions of Richard Bentley, Master of Trinity College, Cambridge 1700-1742, the noted classical scholar remembered for his controversy with Robert Boyle on the epistles of Phalaris.   In 1692, Bentley was appointed to deliver the first series of Boyle lectures, instituted by Robert Boyle in 1691:

> "for providing the Christian religion, against
> notorious Infidels, viz. Atheists, Theists,
> Pagans, Jews and Mahometans, not descending
> lower to any controversies that are among
> Christians themselves:"

Whiston was himself called to deliver the Boyle lectures in 1707.   The Boyle lectureship provided a stage from which men like Richard Bentley, Samuel Clarke and William Derham could become commentators on the Newtonian natural philosophy and explain their thinking about nature and about society.   The cogently reasoned arguments of the Boyle lecturers presented a formidable structure in support of natural religion.[52]

---

* Henry More and Ralph Cudworth had abandoned the
  mechanical system of Descartes in favour of a natural
  order that relied on God's constant providential
  intervention.

In its early years, the Boyle lectureship was dominated by churchmen who were also earnest advocates of Newtonian natural philosophy. Bentley first became interested in Newton's ideas in the summer of 1691, when he consulted Newton about the geometry and astronomy in the Principia.

In this period of Low Church ascendency, the Newtonian natural philosophy as presented in the Boyle lectures was important as a force against atheism and reason was given an ever larger role in theological discussion.

The subject of Richard Bentley's lectures (1692) was The Folly and Unreasonableness of Atheism, and two years later, A Defence of Christianity against the Objection of Infidels (1694). Bentley stressed the final destruction of the world, but foresaw a deluge as the instrument of destruction rather than a universal conflagration. Only through the intervention of God could this constant decay and final destruction of the world be prevented. On the question of how God chose to effect his providential care, Bentley, Samuel Clarke and Whiston believed that the motion which is omnipresent in the Universe arises from God's power.

In 1693, Whiston, at his own wish, was ordained by William Lloyd, Bishop of Lichfield. There was evidently an affectionate relationship between the two men.

> "This truly great and good Bishop, who took me
> to his bosom, and loved me, as I did him most
> sincerely." (53)

It was probably due to the influence of Lloyd* that

---

* William Lloyd (1627-1717) was successively Bishop of St. Asaph, Lichfield and Coventry, and of Worcester. In 1695, Lloyd was a member of the Commission set up by William III to regulate ecclesiastical preferment. He came to be recognised as an almost official interpreter of scripture prophecies. (55)

Whiston developed his interest in the interpretation of scripture prophecies.   Bentley had consulted Lloyd about such matters in his preparation for the Boyle lectures. His interpretation of the prophecies in the Book of Daniel and the Book of Revelation led Lloyd to assure Queen Anne in 1712 that within 4 years the Church of Rome would be destroyed and that the possible date for the destruction of Anti-Christ was 1736.[54]

A volume of correspondence between Lloyd and Whiston, dating from July 1708 to April 1709, relates to the authenticity of the Apostolical Constitutions, the exact time of the Last Supper, and the discernment of 'original sources'.[56]   Lloyd urges Whiston to submit his ideas to the judgment of learned scholars and warns him that he might have to break off his long-standing friendship if he remained adamant in his errors.

There is abundant evidence of Whiston's interest in religious matters during his time at Clare Hall.   A manuscript book of sermons from about 1698 survives, which gives dates and places when these sermons were used.[57] It is not easy to trace the beginnings of his scientific interests.   Whiston's first meeting with Newton was in 1694, and about this time Whiston heard Newton read some lectures in the public schools.   He acknowledges that he did not then have any understanding of Newton's system of natural philosophy but shortly afterwards he applied himself zealously to study the <u>Philosophiae Naturalis Principia Mathematica</u>.   Whiston heard Newton's philosophy greatly praised in a paper read by David Gregory (1661-1710) when

he was professor of mathematics in Edinburgh.[58]   When
answering John Keill's criticisms of his <u>New Theory of the
Earth</u> in 1698, Whiston relates in some detail the various
stages by which he gradually veered from Cartesian philosophy to Newton's mechanical philosophy.[59]   As already
mentioned, Whiston's interest in cosmological theories was
first aroused by Thomas Burnet's <u>Sacred Theory of the Earth</u>.
As regards his training in analytical geometry, Whiston
was probably educated in a similar tradition to Newton
where the works of Descartes and John Wallis were formative
influences.[60]   Just as Newton gave a commentary on the
printed version of Isaac Barrow's (1630-1677) scientific
lectures, a generation later, Whiston was to prepare for the
press a version of Sir Isaac Newton's <u>Arithmetica
Universalis</u>.

Seth Ward (1617-1689) reports the almost incredible
ignorance of mathematics in early seventeenth century
Cambridge.[61]   Perhaps because of the neglect of mathematics, a prerequisite for any understanding of navigation
and its problems, cosmography appears very seldom in the
notebooks of the seventeenth century Cambridge student.

By the time Whiston was an undergraduate at Clare Hall
in 1668 he listed the following works which could come under
the heading of cosmographical studies:

Phil: Cluverji.	<u>Introductio ad Geographiam</u> Amsterdam 1682.
Gul. Bleau.	<u>de Usu Globorum</u> Amsterdam 1668.
John Wilkins	<u>That the Moon may be a World</u> London 1684.

Sr. H. Blunt   <u>Voyage into the Levant</u>
London 1650.

However, John Wallis and Seth Ward studied the science
of mathematics for themselves. Henry Briggs (1556-1630),
educated at St. John's College, Cambridge, improved John
Napier's work on Logarithms. Isaac Barrow occupied the
Lucasian Chair of Mathematics at Cambridge from 1664-1669
and was author of a number of mathematical works.

Whiston's book list in 1688 includes the following
mathematical works:

John Wallis	<u>Operum Mathematica</u>	Pars 1a. Oxford 1657.
Record's <u>Arithmetick</u>		London 1623.
Galtruchij <u>Mathematica</u>		Cantabr. 1668
Hodders <u>Arithmetick</u>		London 1683.
Tacqueti <u>Geometria</u>		Amsterdam 1683
Barrow's <u>Works</u>	(62)	

Whiston would have been a competent mathematician by the
time he left Clare Hall and his interest in Newtonian natural
philosophy would have been stimulated by the popularity of
his first publication <u>A New Theory of the Earth</u> (1696).

Whiston's ill-health obliged him to resign his tutor-
ship and in 1694 he took the place of Richard Laughton as
Chaplain to Bishop John Moore of Norwich while Laughton
started his eminent career as tutor at Clare Hall. During
the years 1694-1697, Whiston may have resided with Bishop
Moore at Norwich but he appears to have spent some time at
Cambridge.

John Moore (1646-1714) appointed Whiston as his

chaplain during the period 1694-1698.　During the brief reign of James II (1685-1688), politics were overshadowed by the religious question.　Many Anglican ecclesiastics believed that the Providence of God would save the country from the triumph of Catholicism.　The 'Glorious Revolution' of 1688 was supported by churchmen like Moore, Lloyd and Tillotson as an act of God's Providence.　Subsequently, loyal supporters of William and Mary rose to power in the church.　This included Low-Churchmen like John Tillotson, Edward Stillingfleet, Simon Patrick, John Moore, Humphrey Prideaux.　John Moore was chaplain to William and Mary before his appointment as Bishop of Norwich.

Whiston was probably influenced by Moore's political and religious views and his scholarly tendencies would have benefited by access to Moore's library.*

1.4　MARRIAGE.　RECTOR　AT　LOWESTOFT

In June 1699, Whiston resigned his fellowship of Clare Hall on the occasion of his marriage to Ruth Antrobus, daughter of George Antrobus, his former headmaster at Tamworth Grammar School.

A touching letter from George Antrobus to Whiston must have been written shortly after the wedding.

---

* The library which Moore collected and retained was famous throughout Europe.　At his death, he had accumulated nearly 29,000 books and 1,790 manuscripts. The library was offered for £8,000 to Lord Oxford in 1714 and on his refusal was sold for £6,000 to George I.　On the instigation of Townshend, George I gave the library to the University of Cambridge.　(63)

Tamworth, June 12th 1699

"Deare Son,

... I thought I could have parted more easily than I find. I promise myself you'll be a kind and loving husband. I have addressed my daughter to be as I have hitherto found her, gentle and yielding.

... She'll not be able to hold out in a disputation but if anything considerable be offer by her, let it be considered in the cool of the day.

... Your truly loving father
George Antrobus." (64)

There is every reason to believe that Whiston's marriage was happy. Four children survived to adult life.

In 1698, Whiston was succeeded in his chaplainship to Bishop Moore by Samuel Clarke. Whiston accepted the living of Lowestoft cum Kessingland in Suffolk (1698-1701).* He devoted himself to the pastoral care of his people, who earned their living by fishing. Parish registers record Whiston's contribution to the relief of disasters caused by fire.(65) Whiston was in Lowestoft for the end of the seventeenth century. John Blaque, the Church Clerk at the end of the century made the following quaint entry on March 24th 1699/1700:

---

* Accounts of Whiston's time at Lowestoft are to be found in the following works:

R. Reeve. MS volume. History of Mutford and Lothingland. Volume 4 relates to Lowestoft. Ipswich and East Suffolk Record Office HD 196: 1110/2/4.

Gillingwater. Manuscript History of Lowestoft and Lothingland (1790). Ipswich and East Suffolk Record Office HD 196: 1110/1/3.

Hutton. Mathematical and Philosophical Dictionary (1795). p.89-94.

All these accounts are based on Whiston's Memoirs. The entry in Hutton's Dictionary is a copy of that in Gillingwater.

> "Kind reader, it was the good pleasure of Almighty
> God to let me finish this jubilee begun by Mr.
> Jacob Rouse, but who shall the next God only
> knows. God preserve the Church of England
> against all its opposers. Amen. Amen."

During the four years that Whiston was at Lowestoft
he kept in touch with his Cambridge colleagues and maintained
his interest in scientific and mathematical matters. He
owns that his sermons were not written but given from short
notes,

> "which gained me time for my other more learned
> studies, without neglecting my cure." (66)

Whilst at Lowestoft, Whiston must have been working on his
Short Chronology of the Old Testament and of the Harmony of
the Four Evangelists (1702). About this time Whiston
corresponded with Rev. Daniel Whitby who was a prebend of
Salisbury Cathedral and author of A Paraphrase and
Exposition of the New Testament.(67)

It may have been their common interest in scriptural
chronology which prompted Newton to choose Whiston as his
deputy in 1702 and to succeed him in the Lucasian chair of
Mathematics at Trinity College.

1.5   WHISTON AS LUCASIAN PROFESSOR

In February 1701 Whiston started his courses of
lectures in the Cambridge schools. His published lectures
span the period 1701 to 1708. Whiston occupied the
Lucasian chair of mathematics from May 1702 until his dismissal from the University in 1710. It was during this
period that the Plumian professorship of Astronomy was

instituted at Cambridge and Roger Cotes was the first to occupy the chair.

In 1710, Uffenbach, a German scholar, visited Cambridge and gives an account of his visit to the coffee houses and his meeting with Whiston.

"As we could find no more we went for a while into the already mentioned coffee house of a Greek. In the morning, you generally meet the learned there, who read the paper and sometimes drink coffee and smoke tobacco. Learned journals are also found here, known by the title of British Apollo and The Athenian Oracle. Everyone, learned or unlearned may find some entertainment in them.

This time we found in the Coffee-house, among other scholars, the famous William Whiston, who by his many singular opinions, and specially on account of the Arianism¼, which he boldly professes, has made himself only too notorious, particularly in his lately published opusculis, wherein he mocks at the Trinity very shamefully; for example, having concluded of these little tracts in nomine Patris, Filii et Spiritus S. he adds at the end of the voluminis as an erratum: dele 'Filii'. I have bought all his things. Another thing that has caused much noise is, he himself baptised his child by a threefold immersion.

He is, it seems, a man of very quick and ardent spirit, tall and spare, with a pointed chin and wears his own hair. In look, he greatly resembles Calvin. He is very fond of speaking and argues with great vehemence." (68)

It would be a mistake to think that these eight years were fully taken up with lecturing duties in natural philosophy. At this time Whiston was actively engaged in his chronological studies. He also interested himself in the organisation of the Charity schools at Cambridge as well as the Society for Promoting Christian Knowledge founded by Thomas Bray. During the years 1700-1711 his four children who survived to adult life were born: Sarah (1700), William (1703), George (1705) and John (1711). Theirs seems

to have been an affectionate family as evidenced by surviving letters. Whiston found time to remember his sons
William and George when meeting his more influential
friends. Sarah's marriage to Samuel Barker seems to have
been an event which benefited the whole family. Thomas
Barker, and later Samuel Barker his son, were both friends
and patrons to Whiston. Lyndon Hall in Rutland seems to
have been a second home to   his wife
Ruth and their three sons. Samuel Barker was a Hebraist
of some note who took an active interest in Whiston's
publications.*

John Whiston established himself at an early age as
a bookseller in Fleet Street, perhaps through the influence
and encouragement of John Senex who had published so many
of Whiston's books and charts. He was one of the

---

* By kind permission of Mr. Bancroft, the Library Super-
intendent of the British Museum, I was allowed to visit
the stacks in search of copies of Whiston's work. The
British Museum holds fifty-five volumes of Whiston's
works which belonged either to Whiston himself or to
his son-in-law, Samuel Barker. These volumes fall
under the following shelf numbers

873. 1. 1-26   26 volumes
873. m. 1-24 made up of 24 volumes
                          (5 sets of Primitive Christ-
                          ianity Revived)
874. 1. 2-6   5 volumes.

Eleven of the fifty-five volumes are inscribed with the
name of Samuel Barker. Most of the remaining forty-
four have manuscript notes by Whiston. Some of these
volumes are revised and corrected in manuscript ready
for a further edition.

873. m. 10-14. 5 volumes of Primitive Christianity
Revived are interesting in that they are printed with
interleaved plain pages on which someone started to
make a Latin translation of the work in manuscript.
This Latin copy is complete as far as page 102 in Volume
2 and has been corrected by Whiston.

first booksellers to issue regular priced catalogues* and enjoyed the nominal distinction of being one of the printers of the votes for the House of Commons.[69]   John Whiston published his books at the sign of Boyle's Head.   He also printed catalogues of Whiston's works as well as notices of subscription for future works.

It was during his period at Cambridge that Whiston first pursued theological ideas about the development of the doctrine of the Trinity.   He always counted his dismissal from Cambridge as a grave injustice and drew no financial help from the University after 1710.   He was replaced by the blind professor, Nicholas Saunderson. Whiston's sons, William and George, were also students at Clare Hall and studied under Saunderson.[70]   After Saunderson's death in 1739 Whiston applied to be reinstated as Lucasian professor but his request was never acknowledged. According to his son John, this letter of application was secreted by Dr. Ashton, the Master of Jesus College, who never showed it to the electors.[71]

---

* The British Museum holds two volumes from 1762 and 1768 which are catalogues of books to be sold by John Whiston.

A Catalogue of many thousand volumes of curious and valuable books including the libraries of Rev. Charles Reynolds, Chancellor of the Diocese of Lincoln and of George Lodington, Esq., by John Whiston 1768 128. i. 17 (5).

A Catalogue of a very choice and curious collection of books which will be sold by John Whiston and Benjamin White.   15th Feb. 1762.   128. i. 11 (2).

John Whiston also bought the library of Edmund Chishull in 1735 and that of Adam Anderson in 1765.

1.6 DEPARTURE FROM CAMBRIDGE. TRIALS FOR HERESY

In August 1708, Whiston presented his Essay on the
Apostolical Constitutions to Dr. Roderick, Vice Chancellor
of the University and asked for his licence to have it
printed at Cambridge.   Whiston eventually published this
essay with several other Sermons and Essays upon Several
Subjects (1709).   Gradually Whiston's heretical views
became known and on 22nd October 1710, he was summoned
before the Vice-Chancellor and a gathering of a majority
of the heads of the Colleges.   He was accused under the
University Statute Ch. 45 De Contumacibus of spreading
doctrines in the University which were contrary to the
established doctrine of the Church of England.   The heretical doctrines were to be found in Sermons and Essays upon
Several Subjects (1709) and the doxology at the end of his
sermon before the governors of the Charity Schools.   On
October 30th 1710, the following sentence was passed by
the Vice Chancellor and heads of eleven Colleges.

> "Whereas it hath been proved before us that
> W. Whiston, Master of Arts, mathematic pro-
> fessor of this University hath asserted and
> spread about in Cambridge, since 19th day of
> April 1709, diverse tenets against the religion
> received and established by public authority in
> this realm contrary to the 45th statute of this
> University; and whereas the said Mr. Whiston
> being required and exhorted by Mr. Vice-
> Chancellor to confess and retract his errors and
> temerity in so doing, did refuse to make any
> such confession and retraction: it is therefore
> agreed by us the Vice-Chancellor and heads of
> colleges, whose names are here underwritten,
> that the said W. Whiston hath incurred the penalty
> of the aforesaid statute and that he be banished
> from this university, according to the tenor of
> the same.   (72)

Whiston and his family moved to London and he was forced to earn a living in whatever way he could. He did not succeed in obtaining a pension from his professorship at Cambridge nor the continued payment of one third of his professorial salary which was allowed in the statutes to a person who,

> "had behaved himself worthily during his continuance in the place, and had not one hundred pounds a year of his own estate."

In 1711, Whiston published his Historical Preface to Primitive Christianity Revived in which he sets out his disagreement with the orthodox doctrine of the Trinity. In June 1711, the upper house of Convocation passed sentence on Whiston's Historical Preface (1711) as containing,

> "... assertions false and heretical, injurious to Our Saviour and the Holy Spirit, repugnant to the Holy Scripture and contrary to the doctrines of the two first general councils and to the liturgy and articles of our church." (73)

This censure was presented to Queen Anne but was mislaid and no further notice was taken of it. Whiston proceeded to publish the full text of Primitive Christianity Revived (November 1711).

In February 1712, the lower house of Convocation applied to the upper house to confirm the censure in the most solemn manner and to censure Whiston in person. Neither of these courses of action was agreed upon. The Convocation could not agree as to the extent of their powers in this case and asked Queen Anne to lay it before the judges. Once more indecision resulted as four of the

twelve judges believed that Convocation had no jurisdiction
to refer to them cases cited before ecclesiastical courts.

Eventually, in 1713, Whiston was prosecuted before
the dean and chapter's court of St. Paul's by Dr. Pelling,
rector of St. Anne's, Soho, in whose jurisdiction Whiston's
house lay.   The commissary, Dr. Hayward, represented that
it was not in his power to degrade a clergyman, this being
the legal punishment incurred by Whiston.   Dr. Pelling
approached the Lord Chancellor and a Court of Delegates
was called in July 1713.   Whiston was called to appear
before the Court of Delegates on October 26th 1713.   On
the day appointed, he did not appear till after the court
had risen and was declared guilty of contempt of court.
Further proceedings were delayed until 1715 when all
heresy was pardoned by an act of grace.   Whiston was
never excommunicated or degraded.*

Whiston's many publications giving accounts of his
dismissal from Cambridge and his trial by Convocation seem
to us today to have thrown him into a difficult position
as regards his future career.   It must be remembered that
this publicity was self-chosen and to some extent aroused
public sympathy for him.

---

* See Appendix III to this chapter for Bibliography of
  publications relating to Whiston's trials for Heresy.

1.7   LIFE AND ACTIVITIES IN LONDON

Information about Whiston's places of residence in
London can be deduced from his correspondence.

1711 - Union Court, near Ely House, Holborn.

1712 - Cross Street, near Hatton Garden.

1726 - Gt. Russell Street over against Montague House.

1730 - His sons George and William were resident at
         Hassel Row, Tottenham Court Road, nr.
         Great Russell Street.

His income from public lectures and publications of
various kinds was supplemented by small annual annuities
from some of his friends;   £20 a year from Sir Joseph Jekyll
and £40 annually from Queen Caroline.   He also had a small
estate near Newmarket in Cambridgeshire which brought in
about £40 per year.

Thomas Barker (1722-1809), son of Samuel Barker and
Sarah Whiston engaged in scientific work and in 1757 published An Account of the Discoveries concerning comets with
the way to find their orbits and some improvements in con-
structing and calculating their places.   It is most likely
that Thomas Barker gained this interest from his grandfather, William Whiston.   Notes on Comets observed between
the years 1723 to 1748 are to be found on the inside covers
of Samuel Barker's copy of Praelectiones Physico-Mathematical
(1710).[74]

Thomas Barker was for many years an assiduous observer
of meteorological phenomena, his principal results being
regularly reported in Philosophical Transactions.   He was
also the author of three religious works of a controversial

nature and this interest too seems to have been derived
from his father, Samuel Barker, and his grandfather,
William Whiston.   Thomas Barker married Ann White, sister
of Gilbert White, the naturalist and author of The Natural
History of Selborne.   At Easter 1737, White was at
Lyndon, visiting his sister Ann and he could possibly have
met Whiston there.

It might at first appear from a list of Whiston's
printed works that he devoted the greater part of his
energy to religious matters.   Closer study of his works,
coupled with an examination of surviving manuscripts con-
siderably alters the picture.   Most biographical accounts
of Whiston have been based on his Memoirs, first published
in 1749, when he was 82.   These Memoirs are largely a
commentary on his printed works taken chronologically and
interspersed with the text of sermons or discourses.

The total impression left is that of a religious
enthusiast who pursued ideas which appear very obscure to
a twentieth century reader.   We should, however, remind
ourselves that in the first half of the eighteenth century
sacred chronology and the chronology of antiquity were
highly respected disciplines.   It seems a fair judgement,
widely expressed by Whiston's critics, that he was stubborn
and unreasonable in the persistent attitude which he maintained with regard to the orthodox doctrines of the Incarnation and the Trinity.   His involvement in these matters
must be seen in the larger context of the great Trinitarian
debate of the eighteenth century.   His fearless attitude
in face of ecclesiastical authority was probably generated

by the authoritarian character of his early religious
training.   Whiston publicly rejoiced in the fact that the
age of true toleration had arrived.   The memory of his own
father's involvement in the restoration of the Church of
England in 1659-1660 was indelibly printed on his mind.
In his writings relating to the literal fulfilment of
Scripture prophecies, Whiston was involved in counteracting
the Deist school of thought which limited the foundations
of religious belief to what was rational.   Whiston's
insistence that God was still active among his people in
fulfilling the Scripture prophecies was an endorsement of
the divine character of God's revelation in the Scriptures.

A study of Whiston's printed works leads me to the
conclusion that he maintained concurrently an active interest
as much in scientific as in religious studies.   Chapter Three
of this work shows that Whiston's attempt to find a solution
to the longitude problem was continuous from 1713 until
about 1744.   The short publications on such methods as
the exploding of guns, the measurement of magnetic dip, the
observation of Jupiter's satellites all presuppose long
periods of experimental work to compile the necessary data.
In this experimental work he kept in contact with other
amateur and professional astronomers.

In 1726 Whiston gave demonstration lectures at
Tunbridge Wells, Bath and Bristol on the Tabernacle of Moses
and of the temples of Solomon, Zorobabel, Herod and
Ezechiel.   He used to carry with him a large model of the
temple and printed illustrations relating to these lectures.
These comprised the <u>Scheme of the Seven Heavens 1727</u> which

was meant as an illustration of the testament of Levi, and
a chart 24" square, which gave a description of the
Tabernacle of Moses and the temples of Jerusalem (1731).
Whiston went again to Tunbridge Wells in 1746 lecturing on
the same subject.   He incurred the displeasure of some of
his contemporaries to such a degree that malicious works
were printed against him.   Conyers Place in 1713 wrote
Heretical Characters Illustrated and Confirmed, a scathing
condemnation of Whiston and his adherence to the doctrines
of Primitive Christianity Revived (1711).   The Scriblerus
Club was started about 1713 and its members included Swift,
Arbuthnot, Parnell, Lord Oxford, Pope and Gay.   In 1731,
the Scriblerians attacked Whiston in Whistoneutes: or
Remarks on Mr. Whiston's Historical Memoirs of the Life of
Dr. Samuel Clarke, written under the pseudonym of Simon
Scriblerus.   A supplement to this work appeared later the
same year by Andrew Scriblerus, entitled Gorgoneicon.
These two publications are a mocking and vicious attack
which ridicule, in vulgar language, Whiston, his writings
and his religious beliefs.

Whiston attracted many comments as a result of his
credulity in the Mary Toft case.   In this, the Guildford
surgeon, John Howard claimed to have delivered nine rabbits
from Mary Toft.   Whiston believed in this monstrous birth
and saw it as a fulfilment of an ancient prophecy.[75]

On the death of Flamsteed in 1719, Whiston considered
applying for the post of Astronomer Royal, but Lord
Chancellor Parker told him that Halley was already nominated
for it.[76]   Whiston never became a Fellow of the Royal

Society. Sir Hans Sloane was to propose him, and Halley
offered to second the motion. Newton is reported to have
said that he would resign from the presidency if Whiston
were to be admitted.[77]

1.8 CORRESPONDENCE AND FAMILY AFFAIRS

Whiston's correspondence indicates a life of interest
and incident. He was a frequent visitor at the Court,
and Queen Caroline is said to have enjoyed his forthright
manner and unequivocal language. Ruth Whiston wrote to her
daughter Sarah from London,

> "August 8th 1730
> "Mr. Whiston is come safe from Windsor and was
> received very kindly at court especially by her
> majesty. She gave him £50 and has told him
> twice she does design something for his son
> though she could not do what he desired last year."[78]

Whiston was not afraid to approach his friends when
seeking employment for his sons William and George. In
March 1726, he wrote to the Right Hon. the Lord Trevor,
Lord Privy Seal.

> "I have two sons of my own, grown up and educated
> in all sorts of academical learning at the
> University of Cambridge under Dr. Laughton of
> Clare Hall; neither of them I believe, much
> inferior to any of their standing either for
> sobriety or diligence: either in the classics
> Greek and Latin or in the several parts of the
> Mathematics and Natural Philosophy ... They are
> now learning to understand and to speak the French
> tongue ... they are capable of being private tutors
> to the sons of nobility, either elsewhere or even
> in the University ... and of what other employment
> they are capable your Lordship will easily judge."[79]

Letters survive from Whiston to his son George offering a remedy for his moods of depression.

London April 21 1736

"... I am heartily concerned at the deep melancholy
your own intenseness of study has brought upon
you...   I well remember that when I have been
long and greatly melancholy in former years, let
me only but get on horseback, and I found myself
better immediately."   (80)

George undertook some tutoring duties, and one of his
pupils was the son of Sir Joseph Jekyll, Master of the
Rolls.

In March 1735, Whiston writes,

"I just now communicated your letter to the Master
of the Rolls; who was highly pleased with the
sensible account of Mr. Jekyll's improvement
under your care."   (81)

In February 1749, only 3 years before his death,
Whiston was once more interceding with his influential
friends for a remunerative post for his son George who was
in poor health.

Whiston wrote to the Speaker of the House of Commons,

Feb. 6th 1749

"... you know what vast pains George took in
learning the Armenian tongue, and much too intense
application George made to perfect the translation
and notes of the principal historian of that
nation, Moses Chorenensis; to such a degree
indeed, in a weak and valetudinary constitution of
body, as has for several years rendered him
incapable of hard study and of most employments
otherwise befitting his learning and capacity.
... you will give me leave to make my address to
yourself ... to assist me and procure some employment or place for my son George, that may be easy
and advantageous to him and that without burdening
him with what his weak nerves will not at present
bear."   (82)

It appears that Speaker Onslow did offer George
Whiston a librarianship in the British Museum but George
begged to be allowed to perform the office by a deputy.(83)

1.9   CLOSING YEARS AND HIS DEATH AT LYNDON HALL

Whiston appears to have kept an interest in all his activities even in the closing years of his life.   During his last four years he seems to have been resident for periods at Lyndon Hall.

On June 23rd 1748, Whiston wrote to the Archbishop of Canterbury pleading that there be a special collect ordered to be said for the Murrain or plague which was at the time destroying horned cattle.

The following February 17th, Whiston was present at the Baptist meeting at Morcot and read the sermon about the Murrain from Exodus IX.   The Archbishop of Canterbury acceded to Whiston's request for a revival of the prayer formerly used on 17th January 1704.   This Collect was first read in Lyndon Church on May 19th 1748.[84]

In August 1748, Whiston was again in London and wrote with great enthusiasm to his son-in-law, Samuel Barker.[85] He was anxious to revive the proposal put forward by Dr. Giles Fletcher, that the Tartars were the descendents of the ten tribes of Israel.*

Whiston was still conducting a course of lectures on religious matters in February 1750.   O.J. De La Fontaine wrote to Whiston thanking him for being allowed to attend the lectures gratis and promising to bring along two or three fee-paying members for the audience on the following Tuesday.[87]

---

* Fletcher's manuscript was found in the study of Sir Francis Nethersoles after his death at Polesworth, Warwick.   Giles Fletcher was an ambassador from Elizabeth I of England to The Emperor of Russia. [86]

Whiston's interest in chronological studies was revived at the adoption of the Gregorian calendar in Great Britain which took place from September 2nd 1752.* Eleven days were lost on the night of September 2nd (old style) resulting in the following day being numbered as September 14th 1752. The civil and ecclesiastical year were both counted from the first day of January instead of from 25th March. The Gentleman's Magazine for March 1751 had details of the implementation of the new calendar. In the same issue, Whiston's letter to the Bishop of London was printed in which he points out that in the Apostolical Constitutions, Christ had already left definite rules for the calculation of the date of Easter. Whiston quotes the relevant portions of this work and thinks that the new paschal tables and Gregorian epacts are more perplexing than the calculations suggested in the Apostolical Constitutions.[88]

Accounts of Whiston which have so far appeared all take quite literally his expressed intention of having left the Church of England to join the Baptists.[89] Whiston left Lyndon Church on Trinity Sunday 1747 and never returned to it again. It is not at all clear whether he did in fact formally separate himself from communion with the Church of England. I am inclined to think that he did not. It was simply his attraction to the Baptist liturgy at Morcot and its resemblance to the worship of the Primitive

---

* Pope Gregory XIII introduced the new calendar into western Europe on October 4th 1582. Its effect was to restore the vernal equinox, which governs the date of Easter to 21st March.

Christian Church which drew him to attend these gatherings.

Whiston died on 22nd August 1752 after a week's illness. The <u>Gentleman's Magazine</u> lists the deaths for July-August 1752. The following entry appears,

"Reverend and learned William Whiston, M.A., sometime professor of Mathematics in the University of Cambridge. He was born, December 9th 1667, admitted a student of Clare Hall in 1686 and chosen a fellow of that College in 1693. In 1700 he was appointed by Sir Isaac Newton to read lectures for him and in 1701 was, by the recommendation of that great philosopher, chosen mathematic professor on his own resignation. In that professorship, he continued till 1711, during which time he so clearly explained the Newtonian philosophy in his mathematical and astronomical lectures, which he then published as to introduce into the University, a noble system, which till then was understood but by few, and those deep geometricians." (90)

The following inscription is found on Whiston's tombstone in the churchyard at Lyndon, Rutland:

His writings shew
his unwearied study
and his extensive knowledge
in various parts of literature.
His sufferings for conscience sake
prove his sincerity.
After a life spent
in piety towards God
and benevolence and
charity towards men,
he rests in hope
through the merits of Christ
of a joyful and blessed resurrection
to eternal life.

CHAPTER 1 - REFERENCES

1. Clark, G.   The Later Stuarts (1660-1714), Second
    Edition (1965), p.20-23.

2. Matthews, A.G.   Calamy Revised (1933), p.xiii,lxi.

3. Bosher, R.S.   The Making of the Restoration Settlement
    (1951), p.138.

4. Dugdale, W.   Short View of the Late Troubles in
    England (1681),p.273.

5. MS. Dr. Williams Library.   Josiah Whiston to Richard
    Baxter.   Index of letters to Richard
    Baxter 6.37.

6. Ibid.

7. Whiston, W. Memoirs of the Life and Writings of Mr.
    William Whiston (1749), p.4.

8. Leicestershire Record Office.   Will of Josiah
    Whiston (1676).

9. Leicestershire Record Office.   Amended Will. Josiah
    Whiston (1685).

10. Ibid.

11. Leicestershire Record Office.   Inventory Josiah
    Whiston (1685).

12. Leach, A.F. Educational Charters and Documents
    1598-1909, Cambridge (1911).

13. Swain, R.H. Correspondence to author, 7th Jan. 1971.

14. Tamworth Grammar School Year Book (1906), p.13.

15. Tyrell, V. (Borough Librarian Tamworth).
    Correspondence with author, 23rd November
    1970.

16. Samuel Blythe's book of Fellows Accounts, Muniment
    Room, Clare College, p.147.

17. Leicestershire Record Office.   Will of William Whiston.

18. Gilmour, John S.L.   Huntia, 2, p.117-137 (1965).
    The Rev. John Laurence.

19. Biographia Britannica (1766), p.1766.

43

20. Field MS. 117B Ruth Whiston to Sarah Whiston, Aug. 8th 1730.

21. Op. cit. (reference 7), p.23.

22. Harrison, W.J. Life in Clare Hall Cambridge 1658-1713 (1958), Heffer, Preface (xi).

23. Ibid. Frontispiece.

24. Op. cit. (reference 7), p.25.

25. Wardale, J.R. Clare College. Letters and Documents (1903), p.115.

26. Op. cit. (reference 22), p.41.

27. Bodleian MS. Eng. misc. d.297.

28. Op. cit. (reference 7), p.35-36.

29. Op. cit. (reference 19), p.1356.

30. Op. cit. (reference 22), p.68-69.

31. Costello, William T., S.J. The Scholastic Curriculum At Seventeenth Century Cambridge, (1958) Harvard U.P., p.13.

32. Ibid.

33. Whiston, W. A Vindication of the New Theory of the Earth from the Exception of Mr. Keill and Others (1698), Preface.

34. Op. cit. (reference 31), p.15.

35. Op. cit. (reference 19), p.1360.

36. Whiston, W. Historical Memoirs of the Life of Dr. Samuel Clarke (1738), p.13.

37. Op. cit. (reference 22), p.75.

38. Op. cit. (reference 7), p.25.

39. Bursar's Records. Muniment Room, Clare College.

40. Op. cit. (reference 22), p.82.

41. Op. cit. (reference 39).

42. Op. cit. (reference 22), p.84.

43. Op. cit. (reference 16), p.147.

44. Op. cit. (reference 7), p.26.

45. Watkins, John. Biographical and Historical
              Dictionary, 3rd edition (1807).

46. Op. cit. (reference 22), p.86-87.

47. Op. cit. (reference 7), p.48.

48. Ibid., pp.82 and 84.

49. Ibid., p.24.

50. Cragg, G.R. The Church and the Age of Reason,
              (1970), Pelican, pp.68-70.

51. Op. cit. (reference 7), pp. 55-58.

52. Jacob, M. Candee, The Church and the Boyle Lectures
              (1969), Ph.D. Thesis, Cornell University,
              p.152.

53. Op. cit. (reference 7), p.33.

54. Op. cit. (reference 52), p.105.

55. Dictionary of National Bibliography (1900), Vol. XXXIII
              pp.436-440.

56. B.M. Add MS. 24197.

57. Bodleian MS. Eng Th. e.45.

58. Op. cit. (reference 7), p.36.

59. Op. cit. (reference 33) Preface.

60. Whiteside, D.T. (Editor), The Mathematical Papers of
              Isaac Newton, Vol. 1, (1967), C.U.P.
              Introd. p.14.

61. Op. cit. (reference 31), p.102.

62. Op. cit. (reference 27).

63. Op. cit. (reference 55) Vol. XXXVIII, pp.360-361.

64. Field MS. No. 91, Leicester Record Office.

65. Parish Registers. St. Margaret's Church, Lowestoft.
              (Consulted by kindness of Mrs. Alice
              Seago, Parish Clerk, November 1970).

66. Op. cit. (reference 7), p.125.

67. Op. cit. (reference 45).

68. Mayor, J.E.B. Cambridge under Queen Anne (1911),
              pp.178-179.

69. Op. cit. (reference 55), Vol. 61, p.9.

70. Bodleian MS. Eng. misc. e.259.
    MS copy of Saunderson's lectures belong-
    ing to George Whiston.

71. A New and General Biographical Dictionary (1761).
    MS note by John Whiston (entry for William
    Whiston).

72. Op. cit. (reference 19), p.4207.

73. Ibid. p.4208.

74. B.M. Shelf Number 874.1.4.

75. Whiston, W. Memoirs of the Life and Writings of
    William Whiston, Second edition (1753),

76. Op. cit. (reference 7), p.296.

77. Op. cit. (reference 19), p.4211.

78. Op. cit. (reference 20).

79. Ibid. No. 115. William Whiston to Lord Trevor.

80. Ibid. No. 123. William Whiston to his son George.

81. Ibid. No. 122. William Whiston to his son George.

82. Ibid. No. 150. William Whiston to the Speaker of the
    House of Commons.

83. Ibid. No. 155. George Whiston to Speaker Onslow.

84. Op. cit. (reference 7), p.405-410.

85. Field MS. No.151. William Whiston to Samuel Barker.
    Aug. 31st 1748.

86. Op. cit. (reference 7), p.576-593.

87. Field MS. No. 151B. De La Fontaine to W. Whiston,
    Feb. 22nd 1750.

88. Gentleman's Magazine, Vol. XXI, p.106-109,

89. Op. cit. (reference 7), p.458-491.

90. Op. cit. (reference 88), Vol. XXII, p.385.

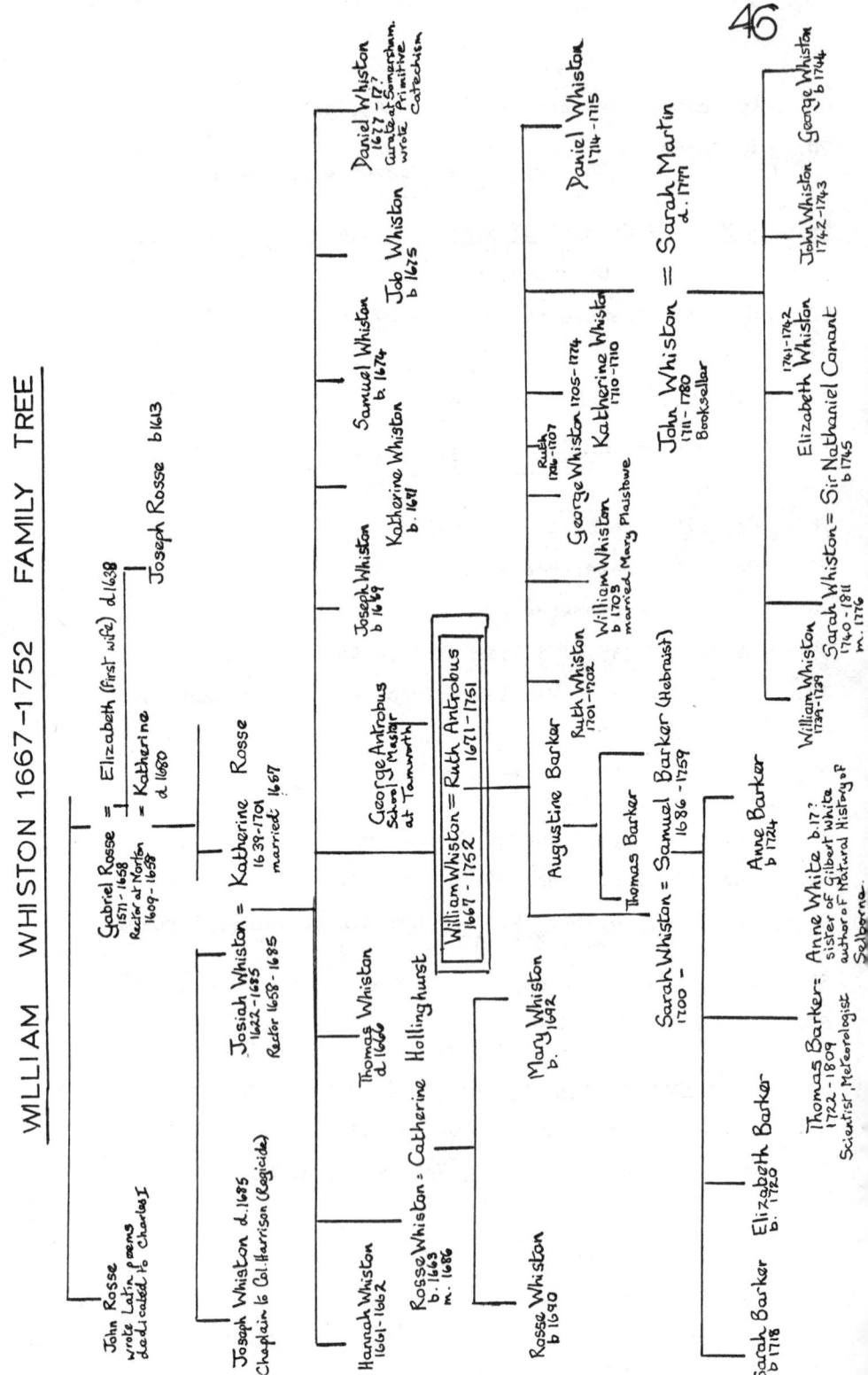

WILLIAM WHISTON 1667-1752 FAMILY TREE

DESCENDANTS OF WHISTON'S GRANDAUGHTER

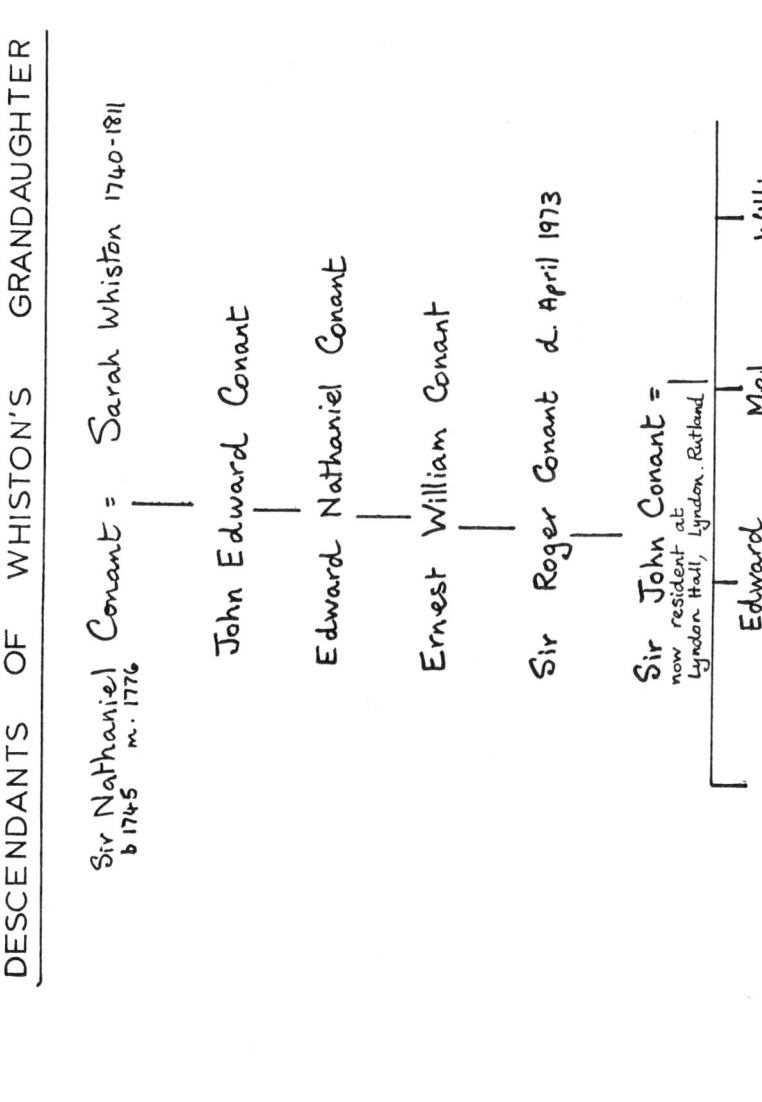

Sir Nathaniel Conant = Sarah Whiston 1740-1811
b 1745 m. 1776

John Edward Conant

Edward Nathaniel Conant

Ernest William Conant

Sir Roger Conant d. April 1973

Sir John Conant =
now resident at
Lyndon Hall, Lyndon. Rutland

Edward   Mel   William

A full family tree — 16 to the
present day — of the Conant family
is in the possession of Sir John Conant
at Lyndon Hall, Lyndon. Rutland

NOTES ON SOURCE MATERIAL FOR FAMILY TREE

1. Memoirs of William Whiston.
    These contain some information about his father,
    and his maternal grandfather but only limited
    information about his own generation of the
    Whiston family.

2. Bodleian MS. Eng. misc. d 297.
    This is Whiston's commonplace book started in his
    boyhood in 1681 when he was about 14 years old.
    It is arranged alphabetically under subjects and
    has 456 numbered pages.
    Pages 6 and 7 contain details of his own brothers
    and sisters, his own family of eight children and
    details of some of his grandchildren.

3. Additional details have been furnished by con-
    sulting the following sources:

    (i)   Parish records in the custody of Rev. J.
          Macdonagh, the Rector (1971) at Holy
          Trinity Church, Norton juxta Twycross,
          Leicester.

    (ii) Bishop's transcripts of the parish
          records of Norton juxta Twycross at
          Archives Department, Leicester Museum,
          New Walk, Leicester.

    (iii) Parish records and inscriptions on
          tombstones in the custody of Rev.
          Anthony Gracie (1970) Rector at
          St. Martin's Church, Lyndon, Rutland.

4. Family tree of the Conant family, now in the
    hands of John Conant of Lyndon Hall, Rutland.

APPENDIX II

**A Catalogue of my Books at Cambridge when
I removed my Chamber, Feb: 1688**

(From the Commonplace book of William Whiston.   Bodleian
MS. Eng. misc. d.297)

In Fol:

Dr. Hammond, On the New Testament.   London (1653).

Sr. Rich. Baker's, Chronicle, with the continuation.
          til 1661.   London (1679).

Sr. Francis Bacon's, Advancement of Learning in English.
          London (1674).

Mr. Clarke's, Mirrour for Saints and Sinners.   London (1657).

Mr. John: Reynold's, Of Murder.   London (1670).

Helvici Theatrum Historicum.   Oxoniae (1651).

Mr. Tho: Fuller's, Holy War and Holy and Profane state.
          Cambridge (1647).

Mr. Leigh's Religion & Learning.   London (1656).

Dr. Hackwel's Apologie of the power and providence of God.
          Oxford (1627).

In Quarto

Dr. Littleton's, Dictionary.   London (1678).

Scapula's Lexicon, Petrum & Jacobum.   Chevet (1621).

Schrevelij Lexicon, cum Robersoni additamentis.
          Cambridge (1676).

Mr. Trellij, Ciceronis Opera omnia.   Genevae (1660).

Mr. Edward Leigh's, Critica sacra.   Heb. London (1642).

Ejusdem Graec: London (1646).

Pezelij, Mellificium Historicum.   Marpurgi (1631).

Mr. John Kettlewel's, Measures of Christian obedience.
          London (1681).

Dr. Stillingfleet's Unreasonableness of Separation.
          London (1681).

Dr. Patrick's, <u>Parable of ye Pilgrim</u>,   London (1665).

Godwins <u>Moses & Aaron</u>.   London (1678).

<u>Roman Antiquities</u>.   London (1690).

Rous, <u>Attick Antiquities</u>.   Oxford (1675).

Dr. Ham.   <u>Pract. Lat.</u> and other works of his.

Dr. Job Venner, <u>Via recta ad vitam longam</u>.   London (1660).

Phil Cluverij, <u>Introductio ad Geographiam</u>.   Amsterdam (1682).

Ich: Magiri <u>Physiologia Peripatetica</u>.   Cantabrigiae (1642).

X Cartesius

<u>Certamen Religiosum</u> between King Charles I and the Marquess
          of Worcester with a vindication of our cause.
          London (1651).

Sr. Kenelm Digby, <u>Of Bodies, and the Soul</u>.   London (1665).

Mr. Streat, <u>Seeming contradictions of scripture reconciled</u>.
          London (1654).

Joh: Wallisij <u>Operum Mathem</u>:   Part 1a. Oxenij. (1657).

Dr. Lightfoot's <u>Harmony</u> in three volumes.   London.

<u>Octavo</u>

Dr. Camber's <u>Companion to the Temple & Altar</u>, on the
          Common prayer.   4 Vol.   London.

<u>Bezae Terti</u>.   Genevae (1580).

<u>Novum Test</u>. cum Stephani Scalligeri Casaubonique notis.
          London (1633).

Gul: Bleau <u>De usu Globorum</u>.   Amsterdam (1668).

Judge Hale's <u>Contemplatione</u>.   London (1682).

Dr. Goodman's <u>Winter evening conference</u> in three parts (1685).

Bp. Taylor's <u>Holy living & dying</u>.   London (1682).

Bp. Wilkin's <u>That the Moon may be a world</u>.   London (1684).

Dr. Patrick on <u>Ecclesiastes & Canticles</u>.   London (1685).

Dr. Fowler's <u>Libertas Evangelica</u>.   London (1680).

Dr. Fowler's <u>Design of Christianity</u>.   London (1676).

Dr. Fowler's <u>Dialogue in Defence of Moderate divines</u>. London (1670).

<u>Psalmi, Proverbia, Ecclesiastes & Canticum</u> ⎫
<u>Canticorum Heb</u>: cum versione interlineari ⎬ Genevae.

Culpeper's <u>English Christian</u> enlarged.   London (1656).

Sr. Charles Wolsely <u>Against Atheism</u>.   London (1669).

Dr. More <u>Against Atheism and Enthusiasm</u> with his Conjectura Cabbalistica.   London.

Homeri <u>Ilias</u>.   Cambridge (1679).

Wadsworth's <u>Immortality of the soul</u>.   London (1670).

Wadsworth's <u>Faith's triumph over the fear of death</u>. London (1670).

Mr. Pool's <u>Nullity of the Romish Faith</u>.   Oxford (1666).

Baxter's <u>Unreasonableness of Infidelity</u>.   London (1655).

Baxter's - <u>32 Directions</u> - London (1657).

Dr. Potter's <u>Answer to Charity mistaken</u>.   Oxford (1633).

Record's <u>Arithmetick</u>.   London (1623).

Boethius <u>Philosophical comfort</u>.   English.   London (1609).

Rogers <u>39 Articles of the Church of England</u>.   Cambridge (1607).

Cambden's <u>Remains</u>.   London (1604).

Brinsley's <u>Grammar School</u>.   London (1612).

Some of <u>Aristophanes & Lucian</u>.   Greek.

R. Bernard   <u>On the Sabbath</u>.

Remigij <u>Daemonolatria</u>.   Lugduni (1595).

<u>Exposition of Doctrine of Church of England</u>.   London (1686).

<u>Conferences with a Jew</u>.   London (1678).

<u>Lucani Pharsalia cum Grotij notis</u>.   Lug. Bat. (1627).

Isocratis <u>Orationes cum Wolfij</u> interpr. a Sam. Crispino (1609).

<u>Officium hominis</u> Lat.   London (1680).

Dr. Hammond's <u>Review</u>.   London (1656).

Ars cogitandi.   London (1682).

Dr. Tillotson's Rule of Faith.   London (1666).

Bercheti Catechismus Lat: Graec: Hanoviae (1628).

Friendly debates, 4 parts 3 volumes.

Stapleton on Juvenal.   London (1673).

Walker on Lillys grammar.   London (1670).

Burgersdicij & Heereboordij Logica

Burgersdicius.   London (1651).

Heerboordius.   Cantabr. (1680).

Horatius cum Bondi notis.   London (1606).

Bishop Sparrow's Rationale.   London (1676).

Tacqueti Geometria.   Amsterdam (1683).

Dr. Patrick's Christian sacrifice.   London (1682).

Dialogi Luciani selecti a du Gardo.   London (1685).

Oxford Grammar.   Oxford (1679).

Dr. Fell's Life of Dr. Hammond.   London (1662).

Seneca Tragaediae.   Oxonij (1679).

Historia ab orbe condito.   Oxford (1669).

Psalms paraphrased.   London (1639).

Eustachij Ethica.   London (1677).

Common prayerbook the best companion.   Oxford (1686).

Walker's Rhetorica.   London (1672).

Dr. Camber's Advice to the Papists.   London (1689).

Bithneri Heb. Gram: London (1675).

Dr. Warmstry's Baptized Turk.   London (1678).

Dr. Wats translations of St. Austins Confessions.
        London (1650).

3 first Cantos of Hudibras.   London (1663).

Mercurius Rusticus and Querela.   Cantabrigiensis (1647).

Walker's Art of teaching.   London (1669).

Galtruchij <u>Mathematica</u>.   Cantabri. (1668).

Mr. Kettlewel <u>on the Sacrament</u>.   London (1683).

<u>Greek commonprayer book</u>.   Cantabr. (1665).

Walker <u>de inventione</u>.   London (1672).

<u>Epitome Gram: Heb</u>: Buxtorfij

Wintertoni <u>Poetae minores Graeci</u>.   Cantabr. (1652).

Galtruchius <u>Poetical History</u>.   London (1678).

Vol. 1 <u>Operum Ciceronis</u> a Lambino Editorum. London (1585).

<u>Court of King James and Charles I</u>.   London (1651).

<u>Maij Historia Parliamenti Angliae</u>.   London.

Clavis <u>Homerica</u>.   London (1638).

<u>Oclandi Anglorum praelia</u> & Nevilli Kettus. London (1582).

Posselij <u>Syntaxis Graeca</u>.   Cantabr. (1640).

<u>Gradus ad Parnassum</u> de est aliquid. Amsterdam (1677).

Golij <u>Ethica</u>.   Argentorati.

Erasmi <u>Colloquia</u>.   London (1639).

Walker's <u>English particles</u>.   London (1676).

Nowel's <u>Catechism Gr: Lat</u>: London (1575).

Graeca <u>Grammatica</u>.   London (1658).

<u>Biblia Tremell</u>: Jun: & Bezae

<u>Greek Testament</u>

<u>Foxes & Firebrands</u>.   Dublin (1683).

Virgilij <u>Poemata cum notis Delphini usui</u>. London (1687).

Tullij <u>de officijs liber</u>.   London (1674).

<u>In Duodecimo</u> etc.

Sr. H. Blunt's <u>Voyage into the Levant</u>.   London (1650).

<u>Juvenalis cum notis Farnabij</u>.   Amsterdam (1668).

Hodder's <u>Arithmetick</u>.   London (1683).

<u>Dying mens words</u>.   London (1682).

Dr. Hammond's <u>Reasonableness of Christian Religion</u>.
London (1650).

Platonis Phaedon. Graece. Col. Agrip. (1579).

Posselij Apophthegmata graecolatinal. London (1652).

Farnabij Index Rhetoricus. London (1676).

Mr. Glanvil Invitation to the Lord's supper. London (1680).

Epitome Justi Lipsij de antiq. Rom. & Fabrilij.
Amsterdam (1657).

Roma velus cum nova collata

Virgilius cum Farnabij notes.

Chamberlain's Introduction into Geography. London (1684).

Mori Ethica. Amsterdam (1679).

Crucij Orationes. Amsterdam (1678).

Herbert's Poems. Cambridge (1641).

Dr. Hall's Cases of Conscience.

Claudianus. Amsterdam (1650).

Plautus. Amsterdam (1652).

Senecae Tragaediae cum Fornatij notis. Amsterdam (1678).

Becani compendium controversiarum. Col. Agrip. (1651).

Tacitus. Amsterdam (1650).

Rous Psalms. London (1643).

Suetonius. Amsterdam (1621).

Common prayer book.

Lipsius in Plinij Panegyricum. Oxford (1662).

Horatius, Juvenalis & Persius (de est folia in 6 satyra
Persij).

Sallustius. Amsterdam (1632).

St. Aug. Confessiones. Col. Agrip. (1637).

Juelli Apologia & Vincentij Livinensis Commonitorium seu
contra hereses. London (1591).

Bible. Oxford (1685).

Pars Senecae operum

Xenophontis Kupo aideias. Geneva (1612).

Sr. F. Bacon's Essays.　London (1668).

Terentij Comaediae

Virgilius.　Amsterdam (1619).

Amesij de Conscientia.　Franckfurt (1635).

Ovidij Opera.　Amsterdam (1619).

Martialis.　Amsterdam (1664).

Parson's Resolution 2nd part.　London (1592).

Fenestella & Pomponius Laetus de magistratibus Romanis.
　　　　　Geneva (1612).

Epitome Romanae historiae.　Raphelengius (?) (1611).

Persius cum Bondi notis.　Amsterdam (1645).

Aesopi Fabulae Graece and Latinae.　London (1657).

Q. Curtius.　London (1672).

Sallustius cum notis.　Oxford (1639).

Lat. Test: (de est aliquid)　Cambridge (1677).

Juvenalis & Persius.　Genevae (1613).

Tullij Epistolae familiares.　London (1584).

Tullij Ad Atticum Brutum & Quintum fratrem.　Lugdum (1585).

Tho: a Kempis De Christo Imitanda
　　　　　Innocentis 3 de contempta mundi　Lugdum
　　　　　　　　　　　　　　　　　　　　　　(1550-1)

Caesaris Commentarij.　Lugdum (1558).

Butler's Rhetorica　　)　Lugdum
　　　　　　　　　　　)　Batavorum
Buchanons Psalms Lat.　)　(1642).

APPENDIX III

Publications Relating to Whiston's Trials for Heresy

Whiston wrote an account of his banishment from
Cambridge and this was first printed at the end of the
Historical Preface to Primitive Christianity Revived 1711.
This was reprinted and sold separately in 1718 under the
title, An Account of Mr. Whiston's Prosecution at and
Banishment from the University of Cambridge.

In 1711 he published An Account of the Convocation's
Proceedings with relation to Mr. Whiston with a postscript
containing a reply to the Considerations on the Historical
Preface and the Premonition to the Reader to which is added
a Supplement to the foregoing Account of Convocation's
Proceedings.   In this publication, Whiston prints all the
letters which he sent to the Archbishop of Canterbury with
the Archbishop's replies, also the address of the Archbishop
and Bishops of the province of Canterbury to the Queen
with Queen Anne's reply.   In the supplement to this work,
Whiston prints the text of,

> The Judgement of the Archbishop and Bishops and
> the Clergy of the Province of Canterbury in
> Convocation assembled, concerning divers asser-
> tions contained in books lately published by
> William Whiston

In a tract of over 150 pages, Whiston writes a defence of
himself under the title Mr. Whiston's Defence of Himself
from Dr. Pelling's Accusation of Heresy (date of 1st edition?)

In 1713 Whiston addressed a final answer against his

persecutors, <u>Reasons for not proceeding against Mr. Whiston</u>
<u>by the Court of Delegates in a letter to the Reverend Dr.</u>
<u>Pelling, rector of St. Ann's Westminster.</u>

The publicity lent to Whiston's prosecution by the
pamphlets mentioned brought others into the controversy.
During the years 1711-1715, several persons wrote a critical
condemnation of Whiston's works.   These included three
publications by Michael Maittaire (1668-1747) the classical
scholar who had come to England at the revocation of the
Edict of Nantes.

Michael Maittaire.   <u>An Essay against Arianism and</u>
<u>some other heresies or a reply to Mr. William Whiston's</u>
<u>Historical Preface and appendix to His Primitive Christ-</u>
<u>ianity Revived</u> (1711).

Michael Maittaire.   <u>Considerations on Mr. Whiston's</u>
<u>Historical Preface being an answer to his plain questions</u>
<u>and other most material passages therein contained</u> (1711).

Michael Maittaire.   <u>The Present Case of Mr. William</u>
<u>Whiston, humbly represented in a letter to the Reverend</u>
<u>the Clergy now Assembled in Convocation</u> (1711).

Another antagonist of Whiston was Styan Thirlby who
wrote three pamphlets.

Styan Thirlby.   <u>An Answer to Mr. Whiston's Seventeen</u>
<u>Suspicions concerning Athanasius in His Historical Preface</u> (1712).

Styan Thirlby.   <u>Calumny, no Conviction</u> (1713).

Styan Thirlby.   <u>A Defence of the Answer to Mr.</u>
<u>Whiston's Suspicions</u> (1713).

Richard Smalbroke was another churchman who opposed
Whiston's ideas.   His 1714 publication, <u>The Pretended</u>

<u>Authority of the Clementine Constitutions Confuted</u>, was one
of his contributions to the debate and in <u>The New Arian
Reproved</u>.

APPENDIX IV

Portraits

1. Oil Painting by Sarah Hoadley (wife of Benjamin
    Hoadley, Bishop of Bangor).
    The original is most probably that in the picture
    gallery at Lyndon Hall.   (See Frontispiece).

2. Copy of the above or possibly a replica of the
    above in the Senior Combination Room at Clare
    College, Cambridge.

3. Another copy of the same work in the National
    Portrait Gallery, London WC24 0HE.   Number 243.
    The portrait of Whiston in this work is a photo-
    graph of the National Portrait Gallery painting.

4. Oil Painting at Lyndon Hall (damaged).
    Whiston is wearing non-clerical dress and holds a
    diagram of comet approaching the sun.   The back-
    ground shows shelves with books.   Legible titles
    read as follows: (1) Newton, (2) Armenian Bible,
    (3) Boerhaave, (4) Historia Armenia.   (See opposite
    page 4.55).

5. An Engraving by George Vertuel published in 1720.
    The frame is adorned with drawings of Norman's
    Dipping Needle (left) and on the right Whiston's
    dipping needle.   The Greek wording below the
    portrait is an extract from the Apostolical
    Constitutions.

6. A portrait by B. Wilson of Whiston as an old man
    appears as frontispiece to the second edition of
    his Memoirs (1753) and is reproduced in Nichol's
    Literary Anecdotes.

7. A "bust" in the National Portrait Gallery.
    This is about 6" high and is in the form of a pipe
    stopper, the carved head at the top is thought to
    represent Whiston.   There is, however, some doubt
    as to whose representation it really is.

CHAPTER TWO

SPECULATIONS IN EARTH HISTORY 1660-1700

WHISTON'S CONTRIBUTION TO THIS DEBATE

2.1 The problems presented by Earth History in the mid 17th century.

2.2 Descartes' Earth in <u>Principia Philosophiae</u> (1644).

2.3 The Importance of the Deluge in Chronological Studies.

2.4 Major English contributions to Earth History.

2.5 Burnet's <u>Sacred Theory of the Earth</u>.

2.6 Woodward's <u>Natural History of the Earth</u>.

2.7 Whiston's <u>New Theory of the Earth</u>.

2.8 The Hydrologic Cycle.

2.9 The Fossil Controversy.

2.10 Reflections on the Origin of Land Forms.

2.11 Terrestrial magnetism, wind systems, plurality of worlds.

2.12 Areas of progress in Earth History during the period 1660-1700.

SPECULATIONS IN EARTH HISTORY, 1660-1700

2.1 THE PROBLEMS PRESENTED BY EARTH HISTORY IN THE MID 17th CENTURY

The forty years between 1660 and 1700 are outstanding for the lively debate carried on in wide areas of natural philosophy, many of which have links with Earth History. The term "Earth History" as used in this context is not to be confined to concepts which today would be classified as "geological". Its use is to be extended to include cosmological theories, studies in biblical chronology and exegesis, natural theology and areas of thought which today would find their place in text-books of astronomy, archaeology and natural history.

The seventeenth century theorists and experimentalists knew no such division of knowledge. Their speculations about the earth, its cosmology and history are in many cases intimately linked with convictions stemming from other areas of natural philosophy and theology. It is rewarding to look at the primary sources with a view to understanding the reasons which prompted the individual men to write these works. In many recent studies of the earth sciences during this period, attention is paid only to one particular aspect, as though attempts were being made to extract from a primary source what today we should call its scientific content. This is to rob the work of its true character, to attribute to the author thought patterns which he would not own.

Speculations dealt with in this chapter have two sources:

(1) Those designed to bring the knowledge of earth history into harmony with the account given in the Book of Genesis and at the same time to incorporate this into a theological system.

(2) Experimental and observational studies of such phenomena as the origin of springs and rivers, the nature and distribution of fossils, the changing topography of the earth.

It sometimes happened that a man made progress in the second area at the same time adhering to conservative views in the first.   John Keill (1671-1721), Savilian professor of astronomy at Oxford, made a worthy contribution to the quantitative study of the water cycle and produced evidence to show that the flow of rivers and streams could be maintained by rainfall.*   At the same time, Keill was unwilling to consider the possibility of attributing the Deluge to natural causes but insisted that it was a direct intervention of God.   Keill in his criticism of Whiston's theory published in 1698 says:

> "... it will also further appear that the Deluge was the immediate work of the Divine power and that no secondary causes without the interposition of Omnipotence could have brought such an event to pass."   (1)

The seventeenth century was a period for widening the

---

* The conservation of the earth's water resources by the processes of evaporation, condensation and precipitation is known as the water cycle.   Studies of the balance maintained by the water cycle must take into account water as ice (solid), water (liquid) in rivers, springs, oceans and water vapour (gas) in the atmosphere.

background against which all cosmological theories were to
be assessed.   Earlier works of importance in this connection include the writings of Nicolaus of Cusa whose
Learned Ignorance (1440) had insisted that though the world
is not infinite, yet it cannot be conceived as finite.
The universe can never be the object of total and precise
knowledge.   It was by the recognition of this necessarily
partial and relative character of our knowledge that Cusa
prepared the way for the concept of an infinite universe.

Later, Giordano Bruno in La Cena de le Ceneri (1584)
wrote:

> "It is certain that it will never be possible
> to find an even half-probable reason why
> there should be a limit to this corporeal
> universe and consequently why the stars which
> are contained in its space should be finite
> in number."   (2)

Bruno probably did not have a great influence on his contemporaries;   only after the great telescopic discoveries
of Galileo did his influence become a factor in the formation of the seventeenth century world view.   Galileo in
his Dialogue on the Two Great World Systems (1632) took no
part in the debate about the infinity of the universe,
though he rejected the concept of a centre of the universe
occupied by the sun or the earth.   Galileo, in fact,
queried whether the centre of the universe existed at all.

2.2 DESCARTES' EARTH AS DESCRIBED IN PRINCIPIA PHILOSOPHIAE (1644)

Descartes clearly formulated the principles of the new cosmology in a mathematical form in his Principia Philosophiae (1644).   Here was found the first theory of the earth which completely ignored the Genesis account and attributed all natural processes, including the evolution of the earth, to mechanical laws.   According to Descartes the earth was once a star and in the course of time the grosser parts of the matter composing it combined to form three layers as shown in Figure 1.

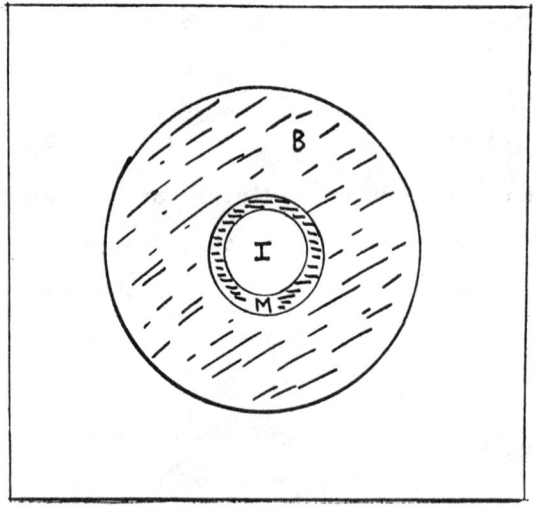

FIGURE 1

The innermost region I, is of the same matter as the sun and moving in a similar fashion.   The second region M is solid and opaque and the third and outermost region B is of loose, varied particles moving freely.

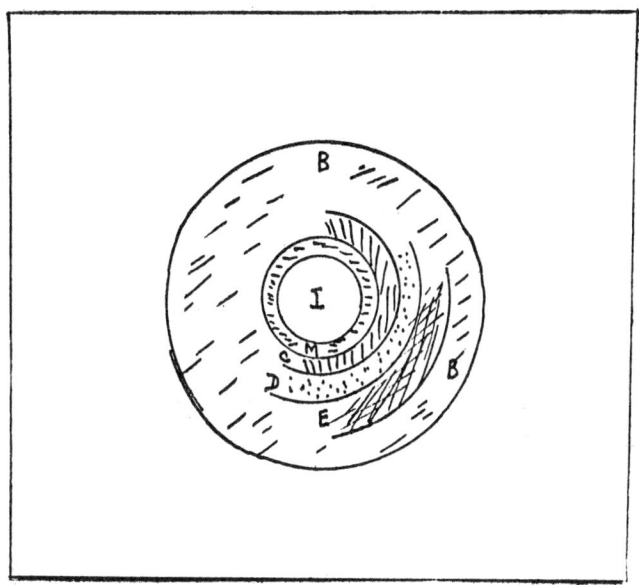

FIGURE 2

As seen in Figure 2, layer B later settles into three different layers. B, the outermost, is the air; the next two layers D and E are the waters, salt and fresh; C, the innermost layer, gives rise to the inner earth and is the source of dry land. There is continual change of the waters into air and <u>vice versa</u> due to agitation by the moving layer of air, B.

Henry More (1614-1687) of Christ's College, Cambridge, exchanged a series of friendly letters with Descartes. This is one of the channels through which Cartesian philosophy and cosmology became accepted in learned circles in England. There is evidence that in

1685, when Whiston was an undergraduate at Clare Hall, Cambridge, one of the principal text books in use was Rohault's Physics, an exposition of Cartesian principles.

2.3   THE IMPORTANCE OF THE DELUGE IN CHRONOLOGICAL STUDIES

The chronological framework into which all English accounts of earth history had to fit was that of James Ussher (1580-1656), Archbishop of Armagh.   His Annales Veteris Testamenti (1650, 1654) calculated that the creation of Heaven and Earth had taken place on the night preceding Sunday, 23rd October, 4004 B.C.   Man appeared on the following Friday; the Flood then took place 1656 years after the Creation.   Noah and his family joined the animals in the Ark on Sunday, 7th December, 2349 B.C.   The waters of the Deluge subsided and the Ark finally came to rest on Mount Ararat on Wednesday, 6th May of the following year. Ussher was a respected scholar and his chronology was the result of prolonged research.   Bishop William Lloyd later revised Ussher's chronology and published the famous inter-leaved bible of 1701.   Whiston was ordained by Lloyd at Lichfield in 1693 and respected his scholarship as a chronologist.   Whiston's chronology as worked out in Book III of his New Theory of the Earth owes much to Lloyd's version of Ussher.

During the seventeenth century the validity of Old Testament chronological studies was never seriously questioned; genealogies and the ages ascribed to the patriarchs were accepted as historical facts.   There was, however, some unease about the claims of ancient civilisations which traced their history back to antediluvian times. An interesting set of correspondence (1699-1707) survives between Richard Allin, of Sidney Sussex College Cambridge, and Whiston. Allin (29th November 1699) discusses the problem of the antiquity of Chinese civilisation.

> "I do not deny that some of Noah's own posterity
> might travel to China after the Deluge, though
> I have no evidence but that the Chinese might
> as well be the posterity of his three sons.
> As to Noah's going to China himself, it cannot
> easily be admitted. For if he went he was
> certainly the head commander of them. But
> the Chinese histories of whose credit we cannot
> reasonably doubt tell us the Yao was King of
> China at that time, who cannot be Noah, because
> he lived in all but 118 years and died in the
> year 2457 (Julian Period) not quite a hundred
> years after the time of the Deluge." (3)

This extract shows the serious scholarship applied to correlation of histories of ancient civilisations with Old Testament chronology.

The Universal Deluge is clearly stated as a fact in Genesis and its date had been accurately fixed by chronological studies; consequently, speculations about earth history have two aspects: the antediluvian world and its characteristics, and the state of the world after the waters had subsided. There were few who dared suggest that perhaps the Deluge was not a catastrophe of immense proportions.

One was Robert Plot (1649-1696), first keeper of the Ashmolean Museum, who in 1677, discussing whether fossils of shell fish were deposited in their present position by the Deluge wrote:

"Not by the Flood in the days of Noah, because
that (and for very good reasons too) seems to
not have been Universal, and at most to have
covered only the continent of Asia and not to
have extended itself to this then uninhabited
Western part of the World."   (4)

Plot cites Edward Stillingfleet's <u>Origines Sacrae</u>
as a source supporting a localised Deluge.   Views such as
that expressed by Plot were among the minority, most writers
considered here accepted the Deluge as a universal catastrophe.   Questions about the causes, duration and results
of the Deluge were asked which could be answered in terms
of observational evidence or mathematical and physical
laws.   The current level of existing knowledge relating to
earth history was organised and recast to answer some of
the following questions.

(i)   What was the source of the waters for the Deluge?

(ii)   How much water was required to submerge the face
             of the earth?

(iii)   What was the extent of the Deluge?

(iv)   Did seismic activity play any role in bringing about
             the catastrophe of the Deluge?

(v)   Were the waters of the Deluge in violent agitation?

(vi)   To what extent did the waters of the Deluge give
             rise to the present topography of the earth?

(vii)   Could the cause of the Deluge be identified with a
             natural cause or was it a direct intervention of
             God?

(viii)   What happened to the waters after the Deluge?

(ix)   If the waters of the Deluge did not originate out-
             side the earth system, how could the same waters

be contained now?

(x)   Could the Deluge explain the present distribution
        of fossils?

(xi)   Were there seasonal variations before the Deluge?

(xii)   Did the position of the earth's axis change at
        the Deluge?

(xiii) How did the new population of plants and animals
        arise after the Deluge?

(xiv)   Was the human population of the earth large before
        the Deluge?

(xv)   Was there a decline in longevity after the Deluge?

2.4 MAJOR ENGLISH CONTRIBUTIONS TO EARTH HISTORY

The following works appeared in close succession:
Thomas Burnet (1635-1715), in Latin, Books I and II of his
Sacred Theory of the Earth in 1681, Books III and IV followed
in 1684 and the whole was later translated into English;
John Ray (1628-1705), The Wisdom of God Manifested in the
Works of Creation (1691), and Three Physico-Theological
Discourses (1693); John Woodward (1661-1728) Essay towards
a Natural History of the Earth (1695), and William Whiston
A New Theory of the Earth (1696).

The publications in this sequence are closely
connected. They drew on common source material and the
works of Woodward and Whiston were critical of the work of
Burnet. Before considering the actual content of these
works it is helpful to examine the different backgrounds

from which each approached his study of the earth's history. Burnet had been a pupil of Tillotson at Clare Hall, Cambridge, had travelled widely on the continent but drew little on observations of natural history to support his theory.   His chief authorities were Scripture and the Classics.

John Ray was a widely travelled natural historian. He had exchanged ideas with Steno in Italy.   The Wisdom of God in the Works of Creation exerted a great influence on scientific and religious thought.   Ray strove to interpret all his data in accordance with the Christian teleological tradition, and developed a general philosophy derived largely from the Cambridge Platonists.   The tradition represented by Ray was carried on by Nehemiah Grew in his Cosmologia Sacra (1701) and reached its climax with Paley's Natural Theology (1802).

John Woodward was a professional physician and was also active in archaeology, ethnology and plant physiology. His interest in earth history was first aroused by the dis-covery of some fossils in the Lias and Oolite near Sherbourne in Gloucestershire.   Woodward was among the first few to appreciate the importance of field observation in this study.   His essay of 1695 was largely based on theories which he could support with field evidence and was meant to correct Burnet's rather fanciful treatment of the Deluge.

Whiston, like Burnet, had been educated in a tradition of Cartesian philosophy and later tried to improve on Burnet's theory by applying Newtonian ideas to the cause of

the Deluge. Whiston's <u>New Theory of the Earth</u> is a useful
analysis of the work of Burnet. He shows by his references that he had followed closely the debate relating to
earth history. Of the four authors grouped here Whiston
was the best equipped to write on earth history from the
point of view of physical causation. He had, however, no
field experience, and his originality lies in his explanation of the Deluge as due to the close approach of a comet.

These four authors cannot be considered in isolation.
The theories of Burnet, Woodward and Whiston drew forth
many critical comments from others interested in the same
problems. Criticisms of an individual theory usually
elicited a further publication in the form of a vindication.
Many of these vindications were inspired with greater
clarity of thought than the original expositions.

Burnet's theory drew about twenty published comments
including the criticisms of the Bishop of Hereford (1685),
Robert Hooke (1689), Erasmus Warren (1690), John Beaumont
(1693), Archibald Lovell (1696), Robert Saint Clair (1697)
and John Keill (1698).

Woodward's theory was strongly criticised by John Ray
and John Arbuthnot (1697). His essay was translated into
Latin, French, Italian and German. The Latin translation
of Woodward's essay was made by the Swiss naturalist
J.J. Scheuchzer and published in Zurich in 1704.

Whiston's most powerful critic was John Keill, and a
pamphlet war ensued between them. Within a few years of
their publication the works of Burnet and Whiston were

translated into German and French.

The theories of Burnet and Whiston, as shown by the
title pages of their works, set out a complete scheme which
embraces not only the creation of the earth and its history
until the time of the Deluge, but went on to deal with the
events which will herald the consummation of the world,
its final conflagration and the emergence of a new earth
where the thousand years rule of Christ can take place.
Thus we see in embryo a cyclic cosmology which is in
accordance with both Reason and Philosophy and is at the
same time an exposition of the account given in Holy
Scripture.

Burnet made a bold attempt to move away from a literal
interpretation of the Scriptures.

> "Tis a dangerous thing to engage the Authority
> of the Scriptures in Disputes about the
> Natural World, in opposition to Reason; lest
> Time, which brings all things to Light, should
> discover that to be evidently false which we
> had made Scripture to assert."   (5)

2.5   BURNET'S SACRED THEORY OF THE EARTH

Burnet considers the changes that must take place in
the original chaos from which the earth was formed.   He
defines this chaos as:

> "a fluid mass, or a mass of all sorts of little
> parts and particles of matter mixed together
> and floating in confusion, one with another."
> (6th edition, p.67).   (6)

The main Themes of Burnet's Theory are most easily
grasped from an examination of diagrams from his work.

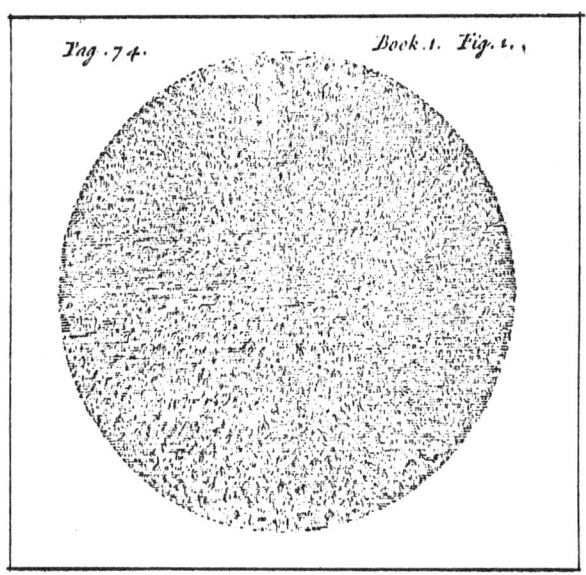

Pag. 74. *Book. 1. Fig. 1.*

FIGURE 3

Figure 3 shows the primitive chaos made up
of small particles floating in confusion.

The remaining diagrams in this chapter, Figures 3 - 15, are
all taken from T. Burnet, <u>Sacred Theory of the Earth</u> (1726)
Sixth edition.

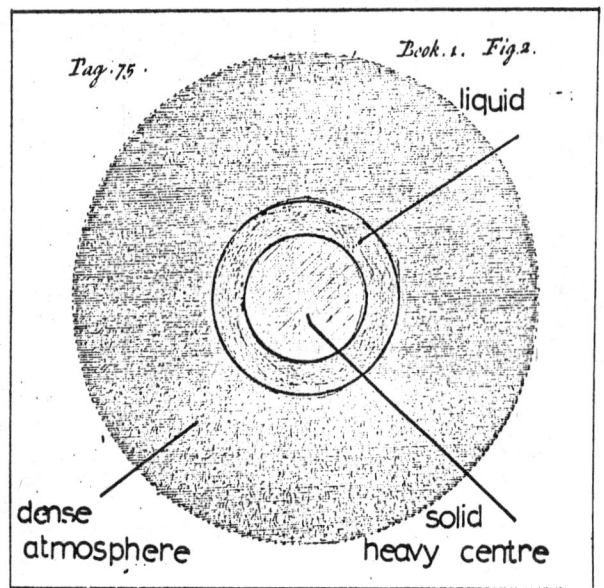

Pag. 75.   *Book. 1. Fig.2.*

liquid

dense atmosphere

solid heavy centre

FIGURE 4

Figure 4 shows the three layers after the
initial settling process, solid heavy centre,
liquids surrounding this and dense atmosphere
containing particles of many kinds.

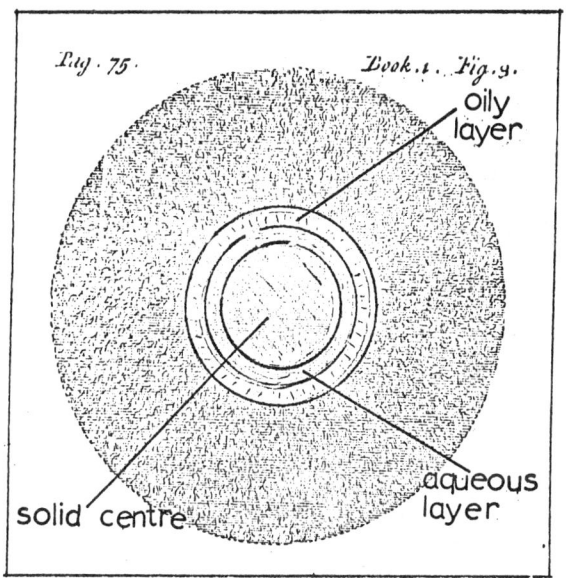

FIGURE 5

Figure 5 shows the separation of the liquids
into two layers, the upper liquid layer being
of an oily consistency floating on the aqueous
layer.

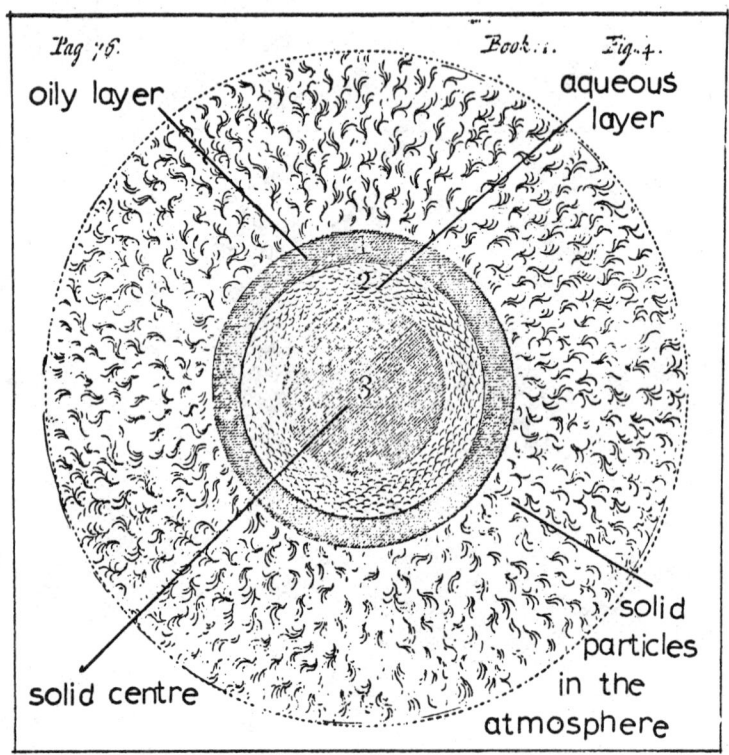

oily layer

aqueous layer

solid particles in the atmosphere

solid centre

FIGURE 6

Figure 6 shows the separation of solid particles from the atmosphere. The lesser and lighter of these would sink more slowly and become entangled with the oily liquor on the face of the deep, where they composed:

> "a certain slime or fat, soft and light earth
> spread on the face of the waters."

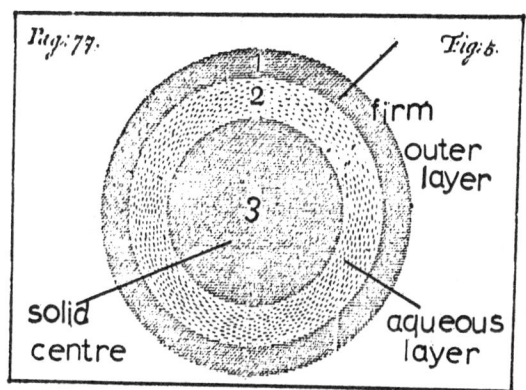

FIGURE 7

Eventually this outer layer absorbed all the solid particles and the oily liquor began to grow more stiff and firm as in Figure 7. This was, according to Burnet:

"The first concretion or firm and consistent substance that rose on the face of the chaos." (7)

He goes on to describe how:

"this body of new concretion ... would become more dry by degrees and of a temper of greater consistency and firmness so as truly to resemble and be fit to make an habitable Earth such as nature intended it for." (8)

These first stages of the earth's history Burnet calls a "Rising World" just as later he called the postdiluvian world a "Fallen World".

He says of this outer crust of the earth:

"What can be a more proper seminary for plants and animals than a soil of this temper and composition? A finer and lighter sort of

earth, mixed with a benign juice, easy and
obedient to the action of the sun, or of what
other causes were employed by the author of
nature for the production of things in the
new-made earth."   (9)

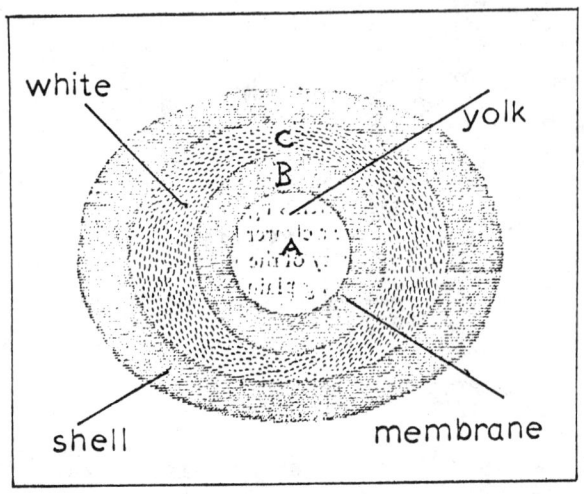

FIGURE   8

Figure 8 represents the earth, flattened at the poles
so that it assumes the form of an egg.   Burnet is
attracted by this representation of an egg-shaped earth as
it appears frequently in the literature of antiquity.
The regions A and B correspond to the central yolk and the
membrane surrounding it.   Burnet leaves open the possibility
that the region A is in fact a central fire.   The region
C, the white of the egg, is the region of the Abyss and
D is the exterior region of the earth.

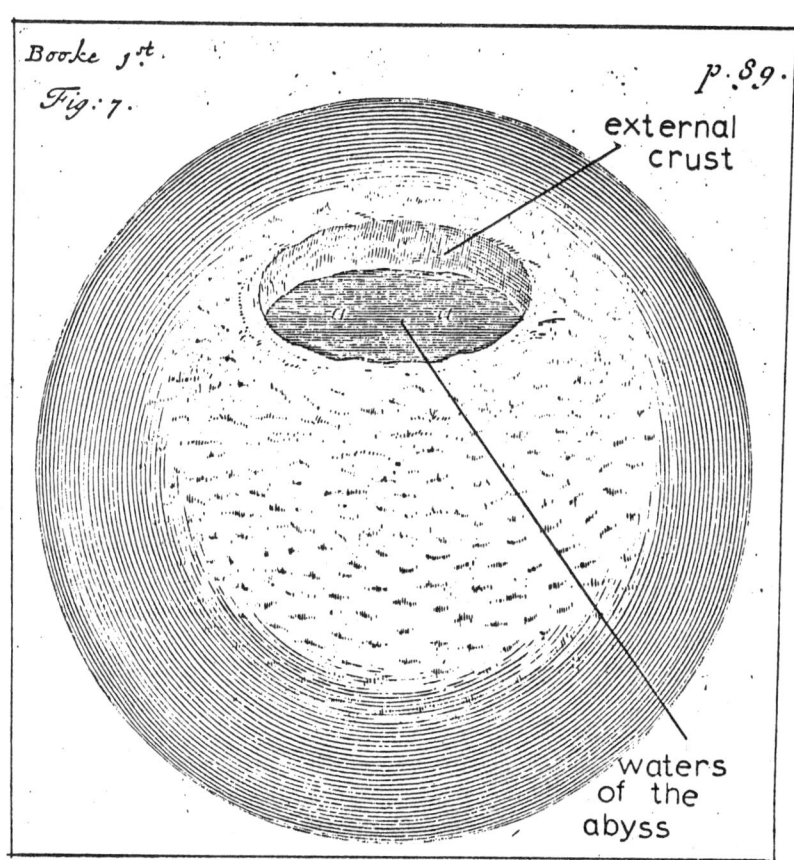

Booke 1st.
Fig: 7.

p. 89.

external crust

waters of the abyss

FIGURE 9

The first earth was smooth and convex and if an
aperture were carved in this smooth surface, the waters of
the abyss would be seen beneath as in Figure 9.

Burnet delights in this first earth:

"In this smooth earth were the first scenes of
the world, and the first generations of mankind,
it had the beauty of youth and blooming nature,
fresh and fruitful, and not a wrinkle, scar or
fracture in all its body, no rocks nor mountains,
no hollow caves, nor gaping channels but even
and uniform all over. And the smoothness of
the earth made the face of the heavens so too;
the air was calm and serene, none of those
tumultuary motions and conflicts of vapours
which the mountains and the winds cause in
ours: 'Twas suited to a golden age and to the
first innocence of nature."   (10)

Burnet was aware of the many accounts of a deluge in
the stories of ancient civilisations, particularly the
epic stories of the Babylonians, Assyrians and Phoenicians.*

Burnet recognises that the great difficulty in explaining the Deluge was to find sufficient water to reach
to the tops of mountains. The Deluge began when the heat
of the sun ruptured the skin of the earth and the force of
the vapours underneath ripped it apart. The frame of the
earth was torn in pieces as by an earthquake and great
portions or fragments of the earth fell into the Abyss in a
state of great disorder.

---

*NOTE. It took a further 190 years of scholarship before
archaeologists were able to trace a written account of the
Babylonian Flood in the Sumero-Akkadian literature. The
eleventh tablet of the Epic of Gilgamesh was found in
Nineveh in 1872 and deciphered by the English Assyriologist
George Smith. This fragment which dates from the 7th
century B.C. contains a version of the Babylonian flood
story. The Gilgamesh Epic was part of the Sumero-Akkadian
literature which lay behind the biblical tradition of the
Flood. As a result of excavations carried out in Ur by
the joint expedition of the British Museum and the University of Pennsylvania Museum a deposit dating from 2900-
2800 B.C. was uncovered. A cut through the floor of the
pit revealed the upper lighter-coloured stratum known as
the "Flood deposit". The darker stratum below consisted
of black organic soil containing potsherds. This "Flood
deposit" is thought to be a relic of the biblical flood and
similar deposits have been found on other sites. (11)

Burnet gives a vivid account of the violence with which the first earth was destroyed:

"The pressure of a great mass of earth falling into the Abyss ... could not but impel the water with so much strength as would carry it up to a great height in the air, and to the top of anything that lay in its way, any eminency, high fragment or new mountain: and then rolling back again, it would sweep down with it whatsoever it rushed upon, woods, buildings, living creatures, and carry them all headlong into the great gulf. Sometimes a mass of water would be quite struck off and separate from the rest and tossed through the air like a flying river." (12)

When the Flood was at its height:

"the Ark was really carried to the tops of the highest mountains, and into places of the clouds, and thrown down again into the deepest gulfs." (13)

As the waters of the deluge retired into their channels,

"God brought a wind upon the waters and the tops of hills became bare and then the lower grounds and plains by degrees." (14)

Burnet acknowledges his debt to Descartes:

"An eminent philosopher of this Age, Monsieur des Cartes, hath made use of the like hypothesis to explain the irregular form of the present earth; though he never dreamed of the Deluge nor thought that first orb built over the Abyss to have been any more than a transient crust and not a real habitable world that lasted for more than sixteen hundred years." (15)

In order to accommodate the waters after the retreat of the Flood, Burnet postulates the existence of large numbers of subterranean cavities, not only those whose orifice appears on the surface of the earth but also many unknown cavities such as miners sometimes come across in the bowels of the earth.[16]

He interprets earthquakes as evidence of the hollowness
of the earth.   He had also given thought to the contemporary problem: the level of the Mediterranean did not rise
even though so many great rivers fed their waters into it.
His answer was that the excess waters discharged into the
Mediterranean were taken by subterranean passages into
the ocean.   In his view the postdiluvian world is an ugly
place,

> "the shores and coasts of the sea are no way
> equal or uniform but go in a line uncertainly
> crooked and broke; indented and jagged as a
> thing torn.   The whole business of the sea-
> channel is but a ruin and in a ruin things
> tumble uncertainly and commonly lie in con-
> fusion."   (17)

Figures 10, 11 and 12 are a series of three diagrams
showing the succession of events which might give rise to
a broken coastline (see following page).

Burnet does not seem to have enjoyed his journey
across some of the mountainous areas of Europe.

> "I have had the experience myself to cross the
> Alps and the Apennine Mountains.   The sight of
> those vast and indigested heaps of stones and
> earth did so deeply strike my fancy, that I was
> not easy until I could give myself some tolerable
> account how that confusion came in nature."   (18)

He gives two maps showing the face of the earth, the
principal mountains and the proportions of land and sea.
The contemplation of these irregular surfaces, so different
from the smooth antediluvian world does not please Burnet.
He reflects:

> "What a rude lump our world is which we are apt
> to dote upon."   (19)

83

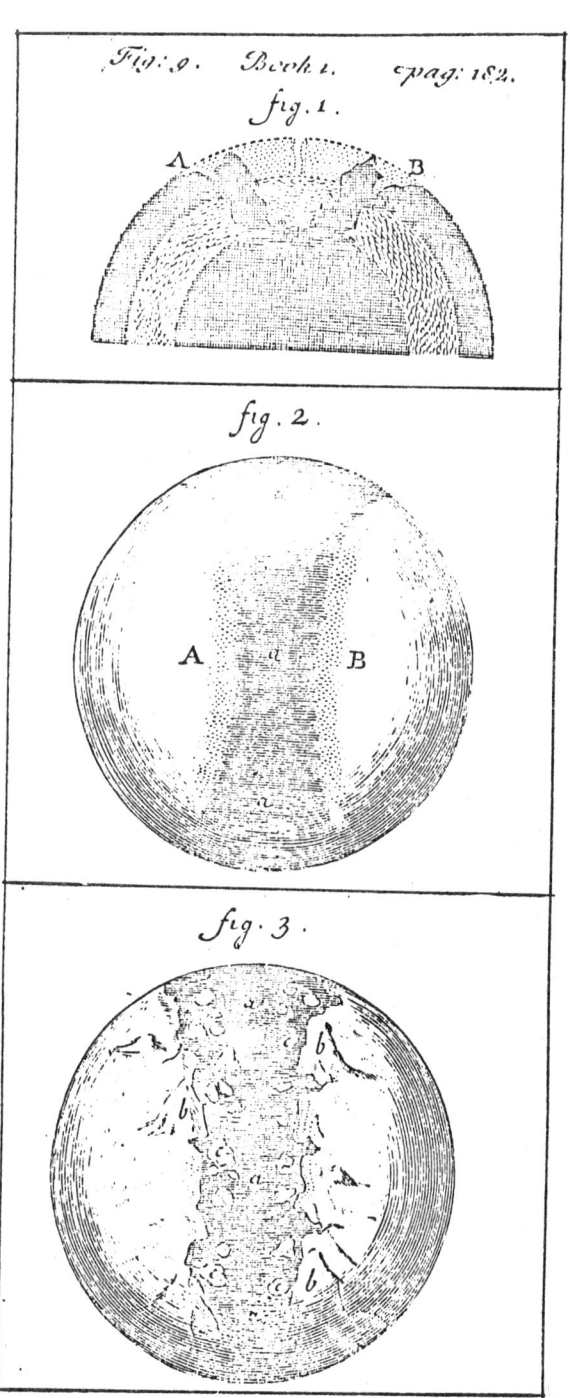

Fig: 9. Book 1. pag: 182.

fig. 1.

fig. 2.

fig. 3.

FIGURE 10

FIGURE 11

FIGURE 12

Burnet found difficulty in explaining the presence
of rivers in the first or antediluvian earth.   To begin
with he tried to explain the circulation of water vapour
(see Figure 13).

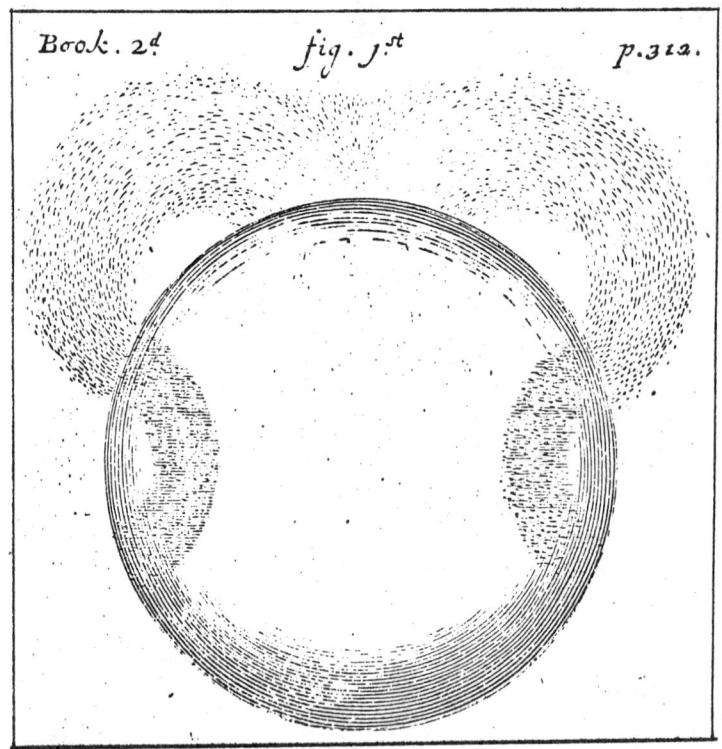

FIGURE   13

The heat of the sun is directed principally to the
middle parts of the earth.   The vapours arise and become
rarefied and follow the course of least resistance, towards

the Poles and colder regions of the earth.*   When the
vapours arrive at those cooler climates, they would be condensed as rain, a continual rain, the same throughout all
times of the year since there was as yet no seasonal
change.[19]

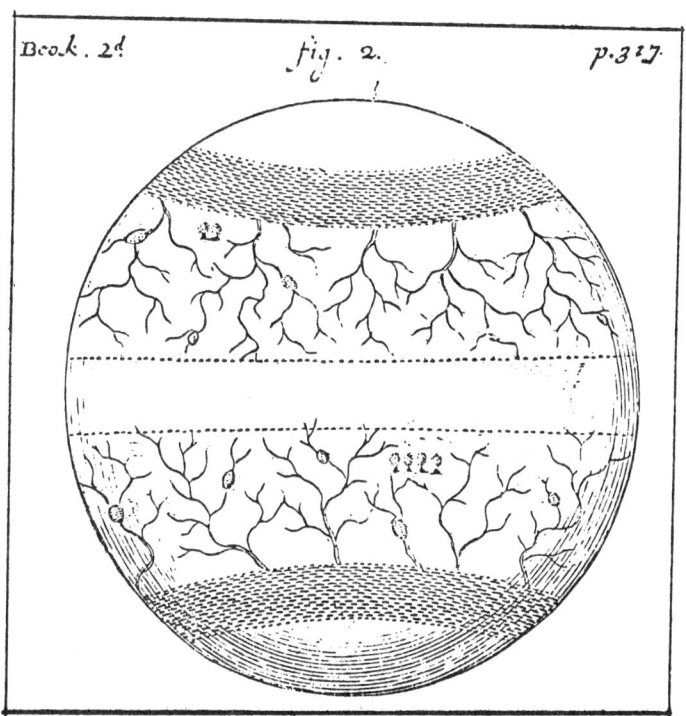

FIGURE 14

---

* Burnet's explanation of the circulation of watery vapours
foreshadows Halley's explanation of the Trade Winds in
1686.

   "But as the cool and dense air, by reason of its
   greater gravity, presses upon the hot and rarified,
   'tis demonstrative that this latter must ascend in
   a continued stream as fast as it rarifies ... so by
   a kind of circulation, the N.E. Trade-Wind below,
   will be attended with a S.W. above, and the S.E.
   with a N.W. wind above."   (21)

Figure 14 indicates that Burnet thought that rivers would originate where the rains fell and would flow towards the equatorial regions of the earth where they would spread out and gradually become absorbed. Burnet's explanation of how the rivers managed to flow over the surface of this smooth primitive earth is not very convincing, since it seems to depict rivers flowing from mouth to source. It was on this subject of the water cycle and the origin of rivers, that Burnet's work was instrumental in provoking more informed discussion.

Books 3 and 4 of Burnet's Sacred Theory are entitled Concerning the Burning of the World and Concerning the New Heavens and the New Earth.

Burnet assigns as natural causes for the final conflagration the burning mountains or volcanoes, also the combustion of oily liquors and pitch or brimstone now dispersed in the earth. He is convinced that the final conflagration will start at Rome, the seat of Antichrist, where the fires are already burning within easy reach of the papal throne!

> "Nature hath saved us the power of kindling a fire in those parts ... since the memory of Man, there have always been subterraneous fires in Italy." (22)

This final conflagration is prophesied in 2 Thessalonians:

> "Out Saviour at his coming in flames of fire shall consume the wicked one, the man of sin, the son of perdition, with the spirit of his mouth, and shall destroy him with the brightness of his presence." (23)

The burning fires on the surface will completely destroy the face of the earth. Earthquakes and sub-terranean eruptions will tear the body and bowels of the earth. The exterior of the earth melted like glass will fill all the subterranean caverns and a smooth regular surface will appear.

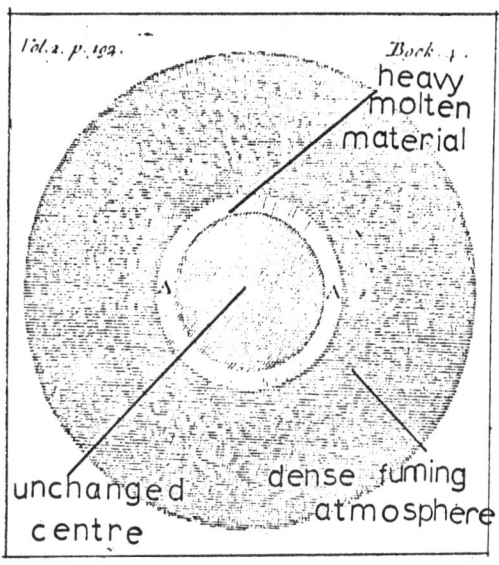

FIGURE 15

Figure 15 shown above illustrates the formation of a new earth.

This outermost region is once more a chaos and in time layers will settle out and the surface of the earth will be regenerated to form a new habitable earth.

".. [it] .. will become such an Earth and of
such a form, as the first paradisaical earth
was."   (24)

This new earth will be inhabited. Burnet does not
think that the conflagration will leave any survivors, but
he is not without ideas whence the new population could
come.

> "Shall we imagine that these new inhabitants
> are a colony wafted over from some neighbouring
> world, as from the Moon or Mercury or some of
> the higher planets?" (25)

Burnet speculates also about the possibility of a
two-fold resurrection with an interval of a thousand years.
Those raised in the first resurrection are those just that
will inhabit the New Heavens and New Earth. The reign of
the Saints will then take place with Christ on earth for a
thousand years. This belief in a Millenium is based on
the Book of Revelation:

> "... And I saw the souls of them that were
> beheaded for the witness of Jesus, and for the
> Word of God ... and they lived and reigned with
> Christ a thousand years."
>
> Revelation, Ch. 20, v.4.

Burnet goes on to discuss the time of the Millenium but
does not come to any definite conclusion. There is no
doubt that Burnet revived interest in the idea of a
Millenium. In the first few decades of the 18th century,
many English prophets proclaimed its imminence. Book 4
of Burnet's work discusses the fate of our world at the end
of the Millenium.

> "I am of opinion that the earth after the last
> day of Judgment will be changed into the nature
> of a Sun, or of a fixed star, and shine like
> them in the firmament.
>
> ... but if planets were once stars as I believe
> they were, their revolution to the same state

Ἐγώ εἰμι τελᾷ ἣ ᾧ.

Τίς κτίσας.   Ἀπὸ καταβολῆς κόσμου

The Sacred Theory of the EARTH

again, in a great circle of time seems to be
according to the methods of Providence, which
loves to recover what was lost or decayed." (26)

The title page of Burnet's work appears in the Plate
opposite and illustrates his cyclic cosmology. The
seven stages already described show the successive steps
through which the earth has already passed or will pass.
These culminate with the drawing on the extreme upper left
when our earth returns to the state of a sun or fixed star
from which it originally came. This explanation of
Burnet's theory in some detail shows the context in which
his deliberations about more strictly geological problems
have to be seen.

John Ray in The Widsom of God in the Works of Creation
shows how all aspects of the physical earth, its motion
and the constitution of its parts, work together under the
plan of Divine Providence to make the world a suitable place
for human habitation.

2.6   WOODWARD'S NATURAL HISTORY OF THE EARTH

Woodward's Essay of 1695 was prompted by his
extensive travels in this country, his study of land forms,
fossils, metals, crystals, minerals and his observations
on springs and rivers. He had acquired an extensive collection of fossils and minerals now housed in the Sedgwick
Museum at Cambridge. He put forward a revised theory of
the Deluge and its effects and includes in his discussion
alterations that have taken place in the globe since the
Deluge. In the preface he boldly states that he will

treat Moses as a historian and put his statements to the
test of observation.   Woodward's work is significant in
that he is one of the first to consider the validity of
his speculations to depend on their correspondence with his
field observations.

He opens his essay with an authoritative assertion of
the organic origin of fossils.   His account of the Deluge
is devised to explain the present distribution of fossils
and particularly their deposition in strata at different
levels.   It reads as though a mixing process of stones,
fossils, plants, animals, teeth and shells took place on an
enormous scale.   The settling out process is described
in graphic terms:

> "This subsidence happened generally and as near
> as possibly could be expected in so great a
> confusion according to the laws of gravity ...
> the said strata, whether of stone, of chalk, of
> coal, of earth, ... lying thus each upon other
> were all originally parallel.
>
> ... that after some time, the strata were
> broken on all sides of the globe: that they
> were dislocated and their situation varied,
> being elevated in some places and depressed
> in others."   (27)

Woodward believed that there is an enormous supply of waters
enclosed in the bowels of the earth.   There is a perpetual
circulation of water in the atmosphere arising from the
surface in the form of vapour and falling again as rain,
dew, hail and snow.   Rivers partly run off into the sea
and partly sink down into the earth.

Woodward is aware also that plants and animals have
a role in the distribution of water.   He recognises that
there must be a source of heat in the interior of the earth

to account for hot springs and volcanoes. He is familiar
with Burnet's theory and does not agree that mountains are
only to be regarded as "rude lumps of earth", but recognises their aesthetic appeal. He insists both on the
universality of the Deluge and on the assistance of a supernatural power in bringing it about. The upper or outermost stratum of earth is in a perpetual state of flux and
change.

> "The upper or outermost stratum of earth: that
> stratum whereon men and other animals tread,
> and vegetables grow, is in a perpetual flux and
> change; this being the common fund and prompt-
> uary that supplies and sends forth matter for
> the formation of bodies on the face of the earth.
> That all animals and particularly mankind, as
> well as vegetables which have had being since
> the creation of the world, derived all the con-
> stituent matter of their bodies successively,
> in all ages out of this fund. That the matter
> which is thus drawn out of this stratum for the
> formation of these bodies is at length laid down
> again in it and restored back into it, upon the
> dissolution of them." (28)

> "... the sea was of the same bigness and capacity
> before the Deluge, but that it was of much the
> same form also, and interwoven with the earth
> in like manner as at this time: that there was
> sea in or near the very same parts of the globe:
> that each sea had its peculiar shells and those
> of the same kinds that now it hath: that there
> was the same diversity of climates." (29)

Woodward argues that the seasons are necessary for
providing the sequence of changes which activate seeds
into growth. The earth must have undergone seasonal change
before the Deluge and therefore its axis must have been
inclined to the plane of the ecliptic.

2.7   WHISTON'S NEW THEORY OF THE EARTH

Whiston's New Theory of the Earth (see plate opposite)
was dedicated to Newton, and attempted to improve on all
the theories so far mentioned.   For two years before the
publication of his work, Whiston had known Newton and had
closely studied his Principia, making an effort to reduce
its principles to a more readily intelligible form.   In
1698 Whiston wrote a Vindication of his New Theory of the
Earth in order to defend it against the remarks of John
Keill.   In the preface to this short work Whiston outlines
the sources of his interest and information in the subject.

Burnet's work was known to Whiston from his entrance
to Cambridge (1687).   A defence of the same work was
Whiston's subject as dissertation for his first degree in
1690.

"This good liking continued with me a great
while after; till my deeper researches into
Mechanical Philosophy and the discoveries contained in Mr. Newton's wonderful book began to
convince me of the indefensibleness of many of
the particulars; and that the whole scheme as
it then lay, could not be justified by the
Principles of sound philosophy, nor did it, upon
better consideration agree with the accounts in
the Holy Scriptures." (30)

In particular, Whiston could not accept Burnet's
explanation of the origin of the seasons because the supposed
shift in inclination of the earth's axis was not explained
in conformity with the laws of mechanics.   He thought he
had a better explanation in the near approach of a comet.
Whiston had followed the discussions excited by the famous
comet of 1680.   He had talked to Richard Bentley, Master

A NEW

THEORY

OF THE

EARTH,

From its ORIGINAL, to the
CONSUMMATION of all Things.

WHEREIN

The CREATION of the World in Six Days,
The Univerſal DELUGE,
And the General CONFLAGRATION,
𝕬𝔰 𝔩𝔞𝔦𝔡 𝔡𝔬𝔴𝔫 𝔦𝔫 𝔱𝔥𝔢 𝔥𝔬𝔩𝔶 𝔖𝔠𝔯𝔦𝔭𝔱𝔲𝔯𝔢𝔰,
Are ſhewn to be perfectly agreeable to
REASON and PHILOSOPHY.

With a large Introductory Diſcourſe concerning the Genuine Nature, Stile, and Extent of the *Moſaick* Hiſtory of
the CREATION.

By *WILLIAM WHISTON*, M. A.
Chaplain to the Right Reverend Father in God,
JOHN Lord Biſhop of *NORWICH*, and
Fellow of *Clare-Hall* in *Cambridge*.

LONDON:
Printed by R. *Roberts*, for *Benj. Tooke* at the
Middle-Temple-Gate in *Fleet-ſtreet*. MDCXCVI.

of Trinity College, about his idea that the waters of the
Deluge might have arisen from the atmosphere of the tail of
a comet.   Whiston read Woodward's Essay (1695) with great
interest and agreed with many of its ideas.   He also took
into consideration the fact that if a comet passed by the
earth at the Deluge, it must alter the earth's annual motion
and period.   Whiston's next step was to undertake elaborate
mathematical calculations to determine whether in fact a
comet did pass by at the appropriate time to cause the
Mosaic Deluge.

Figure 16 on the following page shows Whiston's careful plotting of the orbit of the comet which he considered
gave rise to the Deluge.

Whiston's excitement at his discovery is evident in
his own account,

> "Soon after this I found out another still more
> sure way of discovering the time, nay the very
> day of the beginning of the FLOOD FROM ASTRONOMY." (31)

This is the key point of Whiston's cosmology, the mathematical demonstration of the cause of the flood from
astronomy.   During 1695, Whiston corresponded with Woodward,
discussed a draft of his ideas with Bentley and with Newton.
Whiston evidently got an opinion from Newton which cannot
have been unfavourable, since five years later Newton was
prepared to nominate him as his deputy in the Lucasian
chair of Mathematics at Trinity College.*

---

* Newton was appointed as Master of the Mint in March 1696
and moved to London in the same year.   For the years 1697-
1700 Newton did not give the lectures which were part of his
duty as Lucasian Professor.   In 1701 he appointed Whiston
to undertake his duties at Cambridge and allowed him to
draw the full salary.   In December 1701, Newton resigned
from his professorship and Whiston was appointed as his
successor from May 21st 1702.

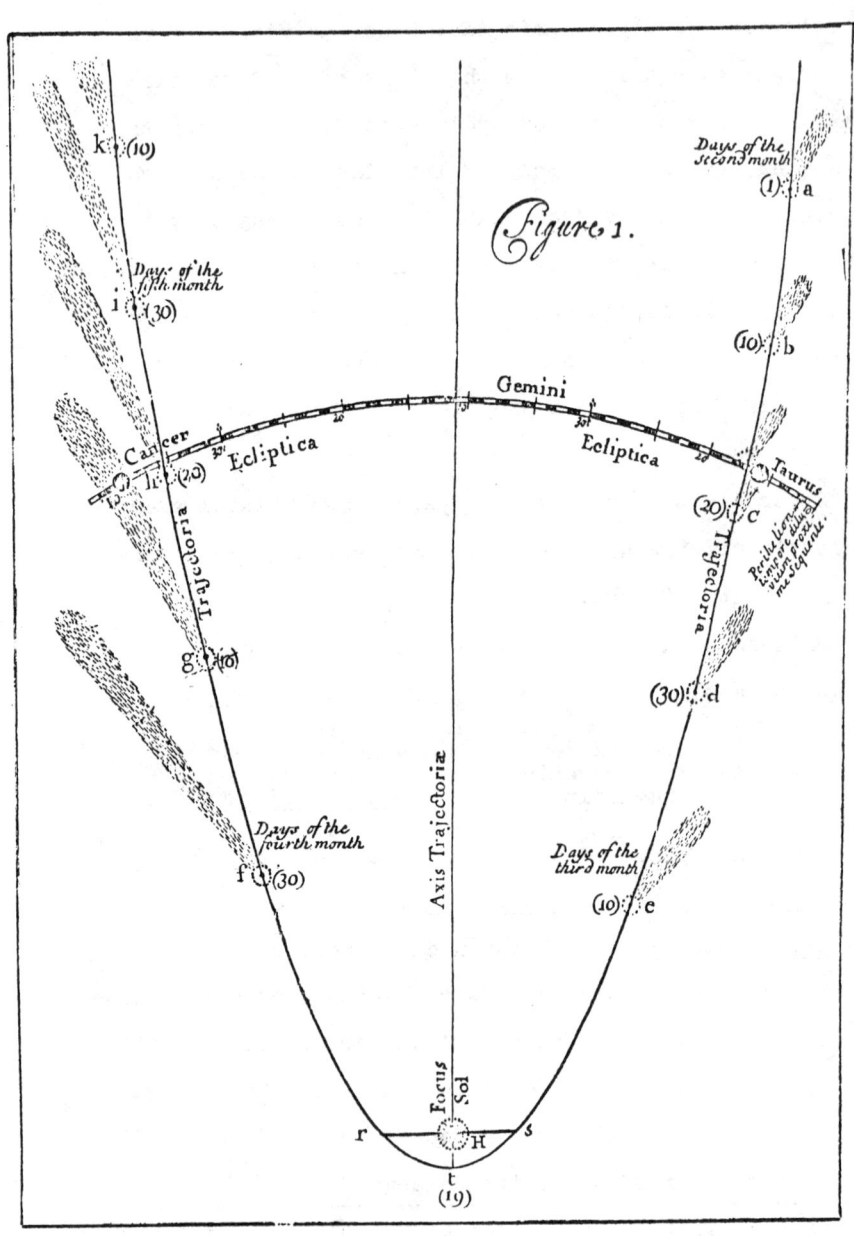

Figure 1.

FIGURE 16

"I brought it to an entire system, and sent it
to Cambridge for Mr. Newton's final review and
correction, which being over, he communicated
to me, I soon brought it into the present form,
and only added that Preliminary Discourse." (32)

Newton had reflected on the atmospheres of comets,
in particular on the comet of 1680. He saw the possibility
of the vapours of the tail of a comet becoming dissipated
and scattered and eventually being attracted towards the
planets by its gravity.* The condensed vapours of comets
could supplement rain water.

"... comets seem to be required that, from their
exhalations and vapours condensed, the wastes of
the planetary fluids spent upon vegetation and
putrefaction, and converted into dry earth, may
be continually supplied and made up." (34)

Newton did not object when Whiston incorporated some
of his ideas on comets into a theory of the earth and its
history.

Whiston was accused by some of having appropriated
his ideas from Halley, but this is unlikely. An entry
in the Journal Book of the Royal Society (May 27th 1696)
recalls that Halley had suggested a similar mechanism as
responsible for initiating the diurnal rotation of the
earth, as early as 1687, in a paper published in the
Philosophical Transactions for that year. Whiston firmly
denied that any of his ideas were derived from Halley.

---

* Professor Harold Urey has recently suggested that
collisions between the earth and "Halley's Comet" type
objects might have been responsible for the dramatic
changes that occurred at the end of geological periods. (33)

"... [I] .. could not possibly make any use of
any notion he [Halley] either proposed or
refuted in this matter, as I am told some persons
have been willing to suggest, and which, if it
had been so, I should freely own." (35)

Whiston presented a copy of his work to the Royal Society,
and Hooke was asked to read it. In July 1696 he commented
on it to the Society.

He drew attention to the fact that Whiston sees the
Mosaic account of Creation as referring only to the formation of the earth out of a preceding chaos and that the
account does not include the formation of the heavens.(36)

Whiston's theory was accepted generally as a serious
work and there is evidence that it became one of the
established texts in philosophical studies to be undertaken
during four years residence at Cambridge.

A closer examination of Whiston's treatment of the
Genesis narrative of Creation reveals a more liberal
attitude to its interpretation than was common in his day.
Whiston states that the words "creating, making or framing"
signify no more than the ordering, disposing, changing or
new modelling of things which already existed.

According to Whiston, before the six days of Creation,
the earth had no diurnal motion. A biblical day and a
year would be of the same duration. This allows six years
for the work of Creation which is a time more appropriate to
the amount of work to be done. Whiston's cosmology can be
divided into seven stages.

1.   The earth had its origin from the atmosphere of a
comet.   The ancient chaos is identified with a central hot

body 7000 or 8000 miles in diameter with a tail or atmosphere about 100,000 miles long.   The comet which gave rise to our earth initially had an elliptical orbit about the Sun.

2.   At the beginning of the Mosaic creation as related in Genesis, the orbit of the comet became circular, the confused atmosphere of the comet subsided and settled downwards.   As yet, the earth had no diurnal motion and its axis was not tilted to the plane of the ecliptic.   This was the condition of the earth at the beginning of the six days of Creation.   The shape of the earth was a perfect sphere.

3.   The six years (biblical days) of Creation followed. At the fall of Man, diurnal rotation began, causing the earth to become an oblate spheroid.   The axis of the earth became inclined to the ecliptic giving rise to the seasons.

4.   At the Deluge, a comet came close to the earth. Owing to gravitational forces, the earth became egg-shaped as shown in Figure 17 on the following page.   A heavier tail effect would be felt at the near pole of approach due to the attraction of the fluids in the abyss, a double tide would strain and stretch the outer layers, which would crack and develop fissures.   The Flood resulted and the waters under the earth were released.   The vapour from the comet, added to the earth's atmosphere, would help to supply the rains associated with the Deluge.   The earth's circular orbit was changed to an elliptical orbit and the length of the year was at this stage increased by 10 days, 1 hour, 30 mins. to become 364 days, 23 hours, 56 mins.

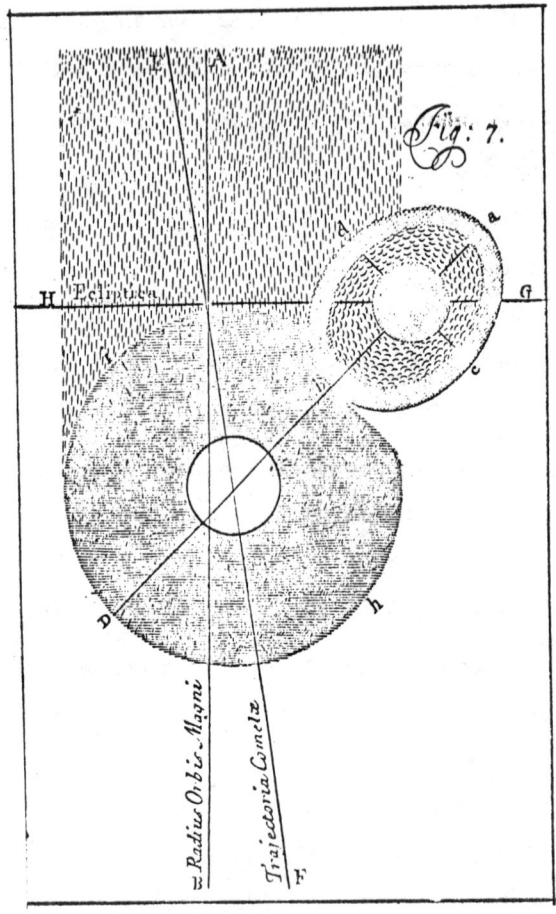

FIGURE 17

5. At the end of this world, the near approach of a comet
will cause over-heating of the earth. The earth will be
oppressed by meteors, exhalations, steams.

6. A new earth will emerge and the resurrection of the
dead will take place. The Saints and Martyrs will live on

earth for a thousand years until the last Judgment.

7.   At the close of the Millenium, the earth "will desert
its present station and no longer be found among the planetary chorus."   Perhaps, through the approach of another
comet, it would revert to its ancient circular orbit or
into some other elliptical orbit, very different from the
present one.

> "... if any comet instead of passing by or gently
> rubbing the earth, lit directly against it, in
> its course either towards or from the sun, it
> must desert its ancient station, and move in
> quite a different elliptic orbit; and so
> probably of a planet become again a comet,
> for the future ages of the world." (37)

Whiston like Burnet had produced a cyclic cosmology,
but now the force effecting change was the influence of
comets.   Whiston acknowledges that he drew on Newton's
reflections on the comet of 1680 for many of his ideas.

In his description of the successive changes in the
earth during the days of creation, Whiston is content to
use an almost identical series of diagrams as appeared in
Burnet's work, but with this difference, that Whiston
regards the centre of the earth as a hot body.   He thinks
that perhaps the first ages of the world were warmer due
to heating effects from the centre.

In 1714, Whiston published The Cause of the Deluge
Demonstrated.   This is a declaration supported by firmer
evidence that the comet of 1680 was in fact the comet which
gave rise to the Deluge.   Whiston consults Halley's Table
of Comets and deduces that the comet of 1680,

"passed before the earth in its annual course
on 17th day of the second month from the
autumnal equinox or November 28th in 2349th
year before the Christian era. The phenomena
of nature and history, and particularly the
mosaic account of the deluge of Noah, which are
not otherwise to be accounted for are exactly
explained." (38)

Whiston quotes extensively from Halley's account of
the comet of 1680 in Gregory's _Astronomy_. Halley traces
this comet back to the year 44 B.C., the year in which
Julius Caesar was murdered. The comet was calculated to
have a period of $575\frac{1}{2}$ years, seven such periods adding up
to 4028 years - exactly the number between 1680 and the
Deluge.

Halley returns to the question of the cause of the
universal deluge in 1724. He had made a study of the rate
of evaporation and condensation of water and was then in a
position to examine the Genesis account of the Flood more
critically.

"Now the rain of forty days and forty nights
will be found to be a very small part of the
cause of such a deluge; for supposing it to
rain all over the globe as much in each day as
it is now found to do in one of the rainiest
counties of England in the whole year, _viz_.
about forty inches of water per diem, forty such
days could cover the whole earth with but about
twenty-two fathoms of water which would only
drown the low-lands next the sea, but the much
greater part would escape."

Halley in 1724 is prepared to modify his views on the
effect of a comet's close approach to the earth.

"What I have advanced I desire may be taken for
no more than the contemplation of the effects
of such a shock as might possibly, and not
improbably, have befallen this lump of earth and
water in times whereof we have no manner of
tradition, as being before the first production

of man, and therefore not knowable, but by
revelation, or else a posteriori, by induction from
a convenient number of experiments or observations,
arguing such an agitation, once or oftener to have
befallen the materials of this globe." (39)

Whiston, like Woodward, agrees that:

"several low countries now bordering on the seas,
might for many years after the Deluge be under
water." (40)

Examples are cited such as Cambridgeshire, Lincolnshire and Holland which still are saturated with the waters
of the Flood.

Whiston tries to explain how the population of the
earth built up rapidly after the Flood, and quotes 400
years as the time required for a country to double its
inhabitants. William Petty and John Graunt had pioneered
a study on population growth which they published in 1662,
Natural and Political Observations made upon the Bills of
Mortality [in the city of London].

An interesting calculation relating to the doubling of
the inhabitants of Breslau survives in a letter from
Richard Allin to Whiston. Allin arrives at 370 years as
the necessary time for doubling the population. In the
same letter, Allin discusses the length of the Antediluvian
year as based on a study of the antediluvian Kings mentioned
in Berossus.* (41)

---

* Berossus, a priest of the great god Marduk in Babylon,
and a scholar, wrote the first account of the Babylonian
Flood in a European language (Greek) about 275 B.C.
(Edmond Sollberger: The Babylonian Legend of the Flood
British Museum, 1971, p.11.)

"That the antediluvian years were just 360 days
long may I think be proved from the Dynasty of
the antediluvian Kings mentioned in Berossus
which is to be seen in Scaligers Thesaurus
Temporum in Graec. The reigns of these Kings
are computed by Sari, each of which are said to
be 3600 years, but because of the ambiguity of
the word arising from the identity of years and
days in the beginning of the world, we may
suppose that a Sarus contains 3600 days, that
is 10 antediluvian years." (42)

Whiston briefly touches on the origin of different
varieties of plants and animals in the post-diluvian world.

"... all seeds, as well of animals, as of plants,
are the immediate workmanship of God." (43)

He thought that the seeds of the post-diluvian animals and
plants were carried by the atmosphere of the comet. Allin
discusses this idea and uses it to explain Noah's drunkenness.

"Noah's drunkenness seems to me a probable argument that there were no vines before the Deluge,
and consequently that the seed of the vine,
among those of other vegetables, came from the
Diluvian comet. If there had been any vine
before the Deluge, it is not likely that the
ante-diluvians would have been ignorant of their
great use." (44)

and in another letter,

"I am persuaded that we have now no remainders
of the Antediluvian vegetables except those whose
seeds were preserved in the Ark, though I will
not deny that many of our postdiluvian vegetables and animals, whose seed came from the
Comet at the Deluge, may be of the same kind
with Antediluvian ones." (45)

Whiston prefaces his New Theory of the Earth with the
assurance that his interpretation of the Mosaic history of
creation in terms of natural phenomena is allowed by the
renowned scripture commentator, Simon Patrick, the Lord

Bishop of Ely.   This approval by Simon Patrick, gave
Whiston's work a degree of acceptance among those who relied
on the authority of the Scriptures.

2.8   PROGRESS   IN   THE   STUDY   OF   THE   HYDROLOGIC   CYCLE

Both Burnet and Whiston relied on the existence of
enormous underground caverns of water to explain the
phenomena of springs and rivers.   Such ideas were largely
derived from Descartes who assumed that the interior of
the earth contained a number of caverns into which water
flowed from the sea through a series of channels.   The
central fire would change the sea water into vapour and
salt would be precipitated, the vapour would condense at
the low temperature of the vault and would be forced upwards and escape through fissures and crevices as springs.

John Keill challenged both Burnet and Whiston on
their views of the hydrologic cycle.   Keill was familiar
with works of John Baptist Riccioli (1598-1671), whose
Geographiae et Hydrographiae Reformata (1661) devotes one
chapter to oceanography and navigation.   Riccioli estimated
the volume of the oceans and included an extensive table
of discharge of rivers of the world.   Keill set out to
show that the flow of streams and rivers could be accounted
for by rainfall.

Between the years 1660 and 1700, significant progress
was made on the measurements which were needed before the
true nature of the water cycle could be proved.

Pierre Perrault (1608-1680), in his De L'Origine des
Fontaines (Paris, 1678) measured the rainfall for three

years, and estimated the run-off of the River Seine.
Results indicated that the total water precipitated as rain
and snow was nearly six times as great as that carried off
by the river. Edmé Mariotte (1620-1682) confirmed Perrault's
work by measuring the rainfall in the neighbourhood of
Paris, and calculated the discharge of the Seine by the
float method. He concluded that subterranean water played
no part in the balance of the water cycle and that springs
were due to rain, snow and other moisture that fell upon
the earth, which

> "filtered through the soil till it met with
> impervious layers in the interior, through which
> it was unable to pass, it therefore continued
> its course along them in an oblique direction
> till it found egress and came out as springs."[46]

Edmond Halley became interested in evaporation when
he was engaged in astronomical work in St. Helena. He
conducted some significant experiments on evaporation of
water at Gresham College in 1693. He estimated the total
quantity of water lost by the Mediterranean in one day by
scaling up evaporation from a pan of water 4" deep and 8"
in diameter. His conservative estimate was that 33
million tons of water would be evaporated each day from
each square degree of the surface. He calculated that
about 5,280 million tons would be lost from the Mediterranean
on a summer's day. His work emphasised the importance of
evaporation in the understanding of the water cycle.

The immense quantity of the waters of the Nile as it
flowed through desert regions raised a huge problem. Few
persons could explain the origin of this quantity of water

without having recourse to the idea of underground water
supplies.   In 1666, Isaccus Vossius published his
De Nile et Aliorium Fluminum Origine dealing with causes
of the inundation of the Nile.   Vossius was another who
condemned the idea of subterranean channels and suggested
that all rivers proceed from the rendezvous of rain waters,
and that rivers ordinarily take their source from hills.

Woodward's idea of subterranean waters exuded as
vapours by the action of a central fire may be derived from
Athanasius Kircher who expressed a similar theory in his
Mundus Subterraneous (Amsterdam, 1664).   However, Woodward
nowhere mentioned Kircher in his Essay (1695).

2.9   THE FOSSIL CONTROVERSY

The controversy about the nature of fossils went
through a crucial period during the years 1660-1700.

Robert Plot states the question very clearly:

> "... which brings me to the great question now
> so much controverted in the world, whether the
> stones we find in the forms of shell-fish, be
> lapides sui generis, naturally produced by some
> extraordinary plastic virtue latent in the earth
> or quarries where they are found, or whether they
> rather owe their form and figuration to the
> shells of the fish they represent, brought to
> the places where they are now found by a Deluge,
> Earthquake, or some other such means, and there
> being filled with mud, clay and petrifying
> juices, have in tract of time been turned into
> stones as we now find them."   (47)

Plot himself, in company with Martin Lister, favours
the view that fossils are not the remains of living
organisms, because of the enormous difficulty in explaining

how these remains came to be in their present location.
He accepts the view expressed in Edward Stillingfleet's
Origines Sacrae (1662) that the flood was not universal,
was not a violent event and at most only covered the
continent of Asia and did not extend to the then uninhabited Western World.   Plot also discounts earthquakes
as responsible for lowering or raising the land masses
because he observes that earthquakes are inconsiderable in
our northern situation.

Plot reserves as one of his chief arguments against
the organic origin of fossils that God's creation does not
include imperfections.   He realised that many fossil
remains bore no resemblance to contemporary forms of life.
Recognition that these fossil remains were once living
organisms would imply that some living creatures did not
successfully survive until the present day and had become
extinct.   This failure of the species to survive and serve
a useful purpose in the creation would be an imperfection
in God's work.

> "... that it seems quite contrary to the infinite
> prudence of nature, which is observable in all
> its works and productions to design everything
> according to an end."   (48)

This sentiment was later echoed by John Ray who towards
the end of his life wavered in his attitude towards the
organic nature of fossils because he too wondered how he
could maintain his argument of the Divine design in nature
if so many products of nature never fulfilled a useful
purpose.

Plot was familiar with Hooke's ideas on fossils

expressed in <u>Micrographia</u> (1665) and with Ray's contributions to <u>Philosophical Transactions</u> in 1673 and 1675 on
the same subject.   Plot had also read Steno's <u>Prodromus</u>
of 1669, Boccone's <u>Recherches et Observations Naturelles</u>,
<u>Lettre 26</u>.   In a series of lectures given to the Royal
Society in 1668 (<u>Discourse on Earthquakes</u>, 1705) Hooke is
convinced that the majority of fossils are the remains of
former marine organisms.

By the time that Woodward wrote his Essay in 1695,
the fossil controversy was practically over.   He is conversant with the ideas of several European treatises such as
the <u>De Glossopetris</u> of Fabio Colonna (1567-1650) (1616) which
maintained that the familiar "tongue-stones" were actually
sharks' teeth.   There was a free interchange of ideas
between English and continental writers.

2.10 <u>REFLECTIONS ON THE ORIGIN OF LAND FORMS</u>

Another aspect of earth history brought into lively
debate was that known today as geomorphology.

Steno in 1669 first illustrated the processes of
deposition of sediments, subterranean erosion as well as
the phenomenon of unconformity.   Hooke's <u>Discourse on
Earthquakes</u>, most of which was written before 1668, contains
all his most important geological ideas.   He regarded the
presence of fossiliferous rocks in the midst of continents
as clear evidence that major changes in the distribution of
land and sea had taken place since the creation.   These
changes he attributed to gigantic earthquakes operating at a

time when subterranean fires burned more fiercely than at present.

Hooke was aware of the role of denudation in moulding topography. He realised the long continued action of sea, rain and rivers is sufficient to destroy mountains and form a featureless plain. The resultant debris was thought to come to rest in the sea and there become consolidated to form new rocks.[49]

Hooke had ideas about a balance in nature and a cyclic progression of events. He believed that as earthquakes raise one portion of the crust, another sinks in compensation; as mountains are worn down, the floors of valleys are built up. He suggested also that portions of the earth's crust might have undergone successive periods of denudation separated from each other by periods of sedimentation.

Samuel Pepys' Diary contains an account of a discussion about the former land bridge between England and France:

> "... and there came Jonas Moore, the mathematician, to us, and there he did by discourse make us fully believe that England and France were once the same continent, by very good arguments, and spoke very many things, not so much to prove the Scripture false as that the time therein is not well computed or understood."[50]

Ray paid a great deal of attention to land forms. In 1673, he gave actual examples of the denudation processes which Steno had described.

> "that the rain doth continually wash down earth from the mountains, and add part of the sea to the firm land is manifest from Lagune or Flats around Venice; and the Camarg or Isle of the

River Rhosne about Aix in Provence in which
we are told that the Watch-Tower had in the
memory of some men been moved forward three
times, so much had been there gained from the
sea." (51)

Ray's friend, Edward Lhwyd, was equally convinced of
the power of denudation. He noted that the steepest
slopes in Wales are to be found on the highest mountains.
He reasoned that the magnitude of rainfall and snowfall
on high land was much greater than on lower land and hence
was a more effective denudation agent.

Richard Bentley wrote in 1693:

"The tops of mountains and hills will be continually washed down by rains, and the channels
of rivers corroded by the streams; and the mud
that is thereby conveyed into the sea will raise
its bottom the higher." (51)

Burnet was impressed by the potency of denudation and
he claimed that some 10,000 years would be ample for the
destruction of all the earth's present relief. Here it
is interesting to reflect that so advanced a concept of
denudation was accepted and understood within the framework
of the Mosaic chronology of six millenia. The result was
that denudation was considered to be a much more powerful
and rapid process than is the case. These observations
on the changing pattern of landforms preceded Hutton's
theory of the earth by about one hundred years. It is not
always acknowledged that so many advances in this field had
been made in the seventeenth century.

2.11 TERRESTRIAL MAGNETISM, WIND SYSTEMS, PLURALITY OF WORLDS

Theories of terrestrial magnetism also underwent development at this period. The publication of William Gilbert's De Magnete in 1600 had resulted in acceptance of the fact that the behaviour of the compass needle is governed by the magnetic effects within the earth. Extensive voyages had established that the needle does not point to the geographical north pole but sets at an angle to the geographic meridian.

In 1641, Athanasius Kircher, in his Magnes, sive de Arte Magnetica had a theory that through the earth run a number of magnetic fibres, the displacement of which form an orderly arrangement. Variation of the compass needle could be explained in terms of displacement of the magnetic fibres.

Hooke (1675) mentioned the theory of a magnetic sphere with an axis and poles distinct from those of the terrestrial globe. Between 1683 and 1714, Halley contributed several papers to the Philosophical Transactions dealing with terrestrial magnetism. His "four-pole" hypothesis attracted a great deal of interest and his world chart of magnetic "variation" published in 1714 was a useful guide to mariners.

In the mid-seventeenth century, there was very little understanding of the world wind systems. Here also Halley made a notable contribution. Through a careful collection of data from all over the world, he compiled (1702) a World Chart showing the prevailing wind systems.

Speculations about the existence of a plurality of worlds were freely expressed at this time. For instance Bishop John Wilkins (1614-1672) <u>The Discovery of a Worlde in the Moon</u> (1638) attracted a wide readership. Burnet and Whiston both speculated about the possible existence of life on other planets. Halley in 1691, perhaps as a flight of fancy, suggested the possibility of other levels of habitation deep down under the surface of the earth.

Fontenelle's (1657-1757) publication <u>Entretiens sur la Pluralité des Mondes</u> (1686) was basically an attempt to popularise the astronomical theories of Descartes. It was widely read in intellectual circles and aroused interest in the possibility of life on other celestial bodies.

All these speculations were known to Thomas Wright of Durham who as early as 1734 tried to explain the relationship of the solar system to the system of stars which form the Milky Way. By 1750 Wright had amplified his picture of a single, simple system of stars to allow of innumerable systems. There was thus a tendency to move away from an earth-centred cosmology to more adventurous theories which embraced many possible world systems.

2.12 <u>AREAS OF PROGRESS IN EARTH SCIENCE DURING THE PERIOD 1660-1700: CONCLUSIONS</u>

During the second half of the seventeenth century:

1.   The practice of explaining natural processes in terms of physical laws was firmly established.

2.   Some cautious steps were taken to disentangle

speculations about earth history from a too literal interpretation of the Creation narrative in Genesis.

3.   Cyclic cosmologies were advanced which promoted the
idea of an earth subject to continual change.

4.   The earth was recognised to be continually subject
to processes of decay and renewal which moulded its
topography.

5.   A better understanding of the water cycle and wind
systems was achieved.

6.   Terrestrial magnetism, though not fully understood,
was associated with fluctuations in the physical system of
the earth.

7.   Speculations about a plurality of worlds paved the
way for a more adventurous cosmological system.

The wide interest generated in England and other
European countries in all these aspects of earth history was
due in no small measure to Burnet, Ray, Woodward and Whiston
and their critics.

CHAPTER 2 - REFERENCES

1. Keill, J.   An Examination of Dr. Burnet's Theory
                of the Earth together with some remarks
                on Mr. Whiston's New Theory of the Earth.
                Oxford, (1698), pp.178-179.

2. Bruno, G.   La Cena de le Ceneri, quoted by
                A. Koyré in From the Closed World to the
                Infinite Universe.   Johns Hopkins Paper-
                back (1968), p.40.

3. Field MS.   Folio 90.   Allin to Whiston, 29th
                November, 1699.

4. Plot, R.   Natural History of Oxfordshire (1677),
                p.112.

5. Burnet, T.   The Sacred Theory of the Earth, 6th Edition,
                (1726). Preface xix.

6. Ibid.   Vol. 1, p.67.

7. Ibid.   Vol. 1, p.77.

8. Ibid.   Vol. 1, p.79.

9. Ibid.   Vol. 1, p.80.

10. Ibid.   Vol. 1, p.89.

11. Sollberger, Edmond.
                The Babylonian Legend of the Flood.
                British Museum.   Third edition (1971).

12. Burnet, T.   Op. cit. (Ref. 5), Vol. 1, p.100-101.

13. Ibid.   Vol. 1, p.134.

14. Ibid.   Vol. 1, p.135-136.

15. Ibid.   Vol. 1, p.153-154.

16. Ibid.   Vol. 1, p.161.

17. Ibid.   Vol. 1, p.176-177.

18. Ibid.   Vol. 2, p.190.

19. Ibid.   Vol. 1, p.205.

20. Ibid.   Vol. 1, p.310-311.

21. <u>Philosophical Transactions</u>. Lowthorp Abridgement
        (1705), Vol.II, p.140.

22. Op. cit. (reference 5), Vol. 2, p.120.

23. Ibid.   Vol. 2, p.118.

24. Ibid.   Vol. 2, p.193.

25. Ibid.   Vol. 2, p.205.

26. Ibid.   Vol. 2, p.313.

27. Woodward, J.   <u>An Essay Towards a Natural History of
        the Earth</u> (1695), p.75-83.

28. Ibid.   p.228.

29. Ibid.   p.253.

30. Whiston, W. <u>A Vindication of the New Theory of the
        Earth from the exceptions of Mr. Keill
        and others.</u> (1698) Preface, p.2.

31. Ibid.   Preface, p.5.

32. Ibid.   Preface, p.7.

33. Urey, H.   <u>Nature</u>, March 2 (242, 32; 1973).

34. Newton, I.   <u>Mathematical Principles of Natural
        Philosophy.</u>   Translated by A. Motte, 1729.
        C.U.P. (1934), p.530.

35. Op. cit. (reference 30)   Preface, p.9.

36. <u>Royal Society Classified Papers</u>, Hooke's Papers No.85.

37. Whiston, W. <u>A New Theory of the Earth.</u>   Second
        edition (1708) p.449.

38. Whiston, W. <u>The Cause of the Deluge Demonstrated</u>
        (1714) p.1.

39. <u>Philosophical Transactions 1720 to 1732, Abridged and
        Disposed under general heads.</u>   Vol. VI.
        p.38-41.

40. Op. cit. (reference 36) p.399.

41. Op. cit. (reference 11) p.11.

42. Field MS. F.92   Allin to Whiston, March 19th 1700.

43. Op. cit. (reference 36) p.335.

44. Op. cit. (reference 41).

45. Op. cit. (reference 3).

46. Althaus, J. The Spas of Europe (1862), quoted in Biswas, A.K., The Hydrologic Cycle Civil Engineering, April 1965.

47. Op. cit. (reference 4) p.111.

48. Op. cit. (reference 4) p.120.

49. Davies, G.L. Proceedings of the Geologists Association (1964), 75 p.493-498. Robert Hooke and his Conception of Earth-History.

50. Pepys, S. Diary, May 23rd 1661. edited by Wheatley, H.B., London (1952) Vol. II, p.38.

51. Ray, J. Observations made on a Journey Through part of the Low Countries, Germany, Italy and France, quoted by Davies, G.L. in Geographical Journal (1673), 132 (2) (1966), p.252-262.

52. Bentley, R. Boyle Lectures. The Confutation of Atheism (1693), quoted in Davies, G.L. op. cit. (reference 49).

CHAPTER THREE

WHISTON AND THE LONGITUDE PROBLEM

The Magnes or Lodestone's Challenge

CHAPTER THREE

WHISTON AND THE LONGITUDE PROBLEM

The Magnes or Loadstone's Challenge

Precedes Robert Norman's <u>Newe Attractive</u> (1581)

Give place ye glittering sparkes,
Ye glimmering Diamonds bright,
Ye Rubies red, and Saphires brave,
Wherein ye most delight.
In breefe yee stones enricht,
And burnisht all with gold,
Set forth in Lapidaries shops,
For Jewels to be sold.
Give place, give place I say,
Your beautie, gleame and glee,
Is all the vertue for the which,
Accepted so you bee.
Magnes, the Loadstone I,
Your printed sheaths defie,
Without my helpe, in Indian Seas
the best of you might lye.
I guide the Pilots course,
his helping hand I am,
The Mariners delights in me,
So doth the Marchant man.
My vertue lies unknowne,
my secrets hidden are,
By me the Court and Common-weale,
are pleasured very farre.
No ship could sail on seas,
Her course to run aright,
Nor compasse show the ready way,
were Magnes not of might.
Blush then and blemish all,
bequeath to me thats due,
Your seates in golde, your prive in plate,
which Jewellers doo renue.
Its I, its I alone,
whom you usurpe upon,
Magnes my name, the Loadstone cald,
The prince of stones alone.
If this you can denie
then seeme to make reply,
And let the painfull seaman judge,
the which of us doth lye.

This poem illustrates the magical faith placed in the loadstone as well as its proven practical value to mariners and merchants.

The use of the compass needle in finding direction
was first known in western Europe in the twelfth century.
The later discoveries of the variation (declination) and
inclination (dip) of the compass needle continued to inspire
the hope that the loadstone would also be the key to solving
the longitude problem.*

---

* Throughout this work, magnetic terminology is adopted
as follows.

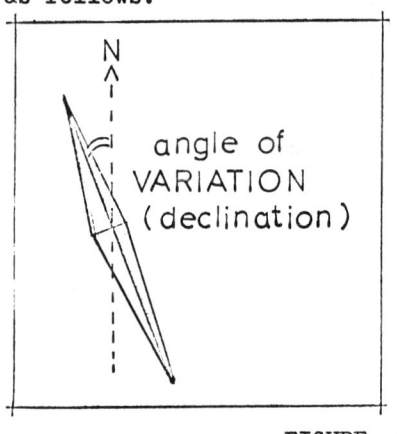

FIGURE 1

Variation is to be
understood in its
earlier usage as the
angle between the
geographic and mag-
netic north.   The
modern equivalent
term is declination.

FIGURE 2

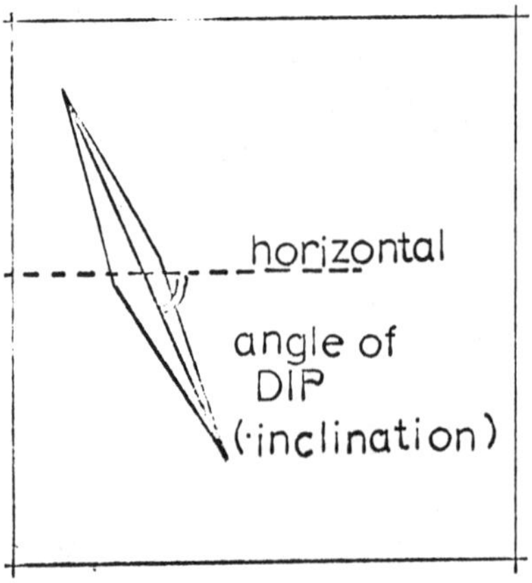

Dip is the angle
between the horizont-
al and a compass
needle when the latter
is pivoted at its
centre of gravity in
a vertical plane and
placed in the mag-
netic meridian. This
quantity is occasion-
ally referred to as
the angle of
inclination.

Attempts to find longitude by charting isoclinic and isogonic lines continued into the eighteenth century. It therefore seems appropriate to preface this chapter by an outline of the principal ideas concerning terrestrial magnetism.

3.1 EARLY HISTORY OF THE LODESTONE IN WESTERN EUROPE

The year 1187 A.D. is now generally accepted as the date of the earliest mention in literature of the employment of the magnetic compass in navigation. Alexander Neckham's description in De Utensilibus (1187) does not make clear whether he refers to the use of a pivoted needle or to a floating magnetic needle. The latter sense seems more likely. In De Utensilibus we read

> "Therefore he who wants to have a well-equipped ship ... let him also have a needle placed under a dart; for the needle will rotate and revolve until the point of the dart looks towards the East, and thus sailors perceive in which direction they ought to go, when the Little Bear is hidden in disturbed weather." (1)

A later compass is known in which, instead of a card there is a little bird whose outstretched beak is made to point to the East by means of a magnetic needle fastened across the wings. The idea of Neckham's compass having indicated East instead of North is not, therefore, altogether strange.(2)

Soon after its mention by Neckham, several writers refer to the compass in one or other of its primitive forms, but none of them appears to know of any departure from the rule then accepted, that the magnetic needle pointed to the north or pole star. The earliest reference to anything of this kind is to be found in the Opus Minus of Roger Bacon (1266). Bacon clearly did not recognise declination as a universal phenomenon.[3] Contemporaneous with Roger Bacon, we have the Epistola of Petrus Peregrinus of Maricourt (1269). The full title of this work is Epistola Petri Peregrini de Maricourt ad Sygerum de Foucancourt militem, de magnete.[4]

There is no evidence that Peter Peregrinus knew of magnetic variation, though it may have been observed by Venetian navigators early in the fifteenth century, for example the chart of Andrea Bianco dated 1436.[5] Usually, however, it is believed that it was first observed by Christopher Columbus during his voyage to the West Indies in 1492.

3.2   THE LONGITUDE PROBLEM: DEVELOPMENTS IN THE SIXTEENTH CENTURY

Great advances in position finding by astronomy were made by the Portuguese in the second half of the fifteenth century, thus making practicable long voyages out of sight of land. Latitude could be found to within 30 miles by observation of the meridian altitudes of the sun and certain stars. By the sixteenth century, charts, navigational almanacs with nautical tables and methods of latitude calculation had appeared. By the early years of the seventeenth century, the use of Ephemerides was becoming usual among skilful seamen. John Tapp's Seaman's Kalendar first appeared in 1601 and gave tables for the years 1602-5. The

tables included the sun's right ascension and declination
and the day and hour of the new moons; the book also
included the catalogue of the fixed stars and the tables
of longitudes were both revised and enlarged.[6]

From the late fifteenth century onwards, the great
need was a method for determining the longitude at sea.
This problem is inseparably associated with the rotation
of the earth on its axis, and thus with the measurement of
time.   Various astronomical possibilities of finding the
time corresponding to a fixed meridian were investigated.
As early as 1514, the German astronomer Johan Werner had
proposed the method of lunar distances.   Werner recommended measuring the distance between the moon and a star
at an observatory at a given time so that an observer who
made the same observation elsewhere, by comparing the local
time of his observations with observatory time could
determine the difference and so his longitude.   The moon
moves across the background of the stars at a rate of about
one minute of arc in two minutes of time.   The resolving
power of the unaided human eye is about 1 minute of arc,
consequently it is impossible to determine terrestrial
longitude using lunar measurements made with this order of
error more accurately than within about 30 miles.   Even
this accuracy pre-supposes that the positions of the fixed
stars and the motions of the moon are accurately known.
In the sixteenth century this was not so, for the typical
observational error was about 5 minutes of arc.   The
method of lunar distances was again advocated by Peter Apian

in 1524, Gemma Frisius in 1530 and Pedro Nunez in 1560.*

The obvious solution of a clock was mentioned by the Flemish scientist, Gemma Frisius in 1530. He recommended that the time at the port of departure should be carried continuously in the ship by means of a mechanical watch, so that when the local time of the ship was found by calculation and observation of the sun's altitude, the difference between the time at the port of departure and the time where the ship was could be found. This would give the navigator his longitude.

Two horological developments, probably made in the fifteenth century made this proposal feasible, the substitution of a driving spring in place of driving weights and the invention of the fusee to ensure that the force exerted by the main-spring should be constant. From the middle of the sixteenth century a number of progressive navigators did carry watches with them in attempts to use them to determine longitude.[8]**

In 1582, Claude de Bossière's translation of Gemma Frisius' work on cosmography was published, Les Principes d'astronomie et Cosmographie; avec L'usage du Globe (Paris, 1582).

A selection of errors in longitude given by Bossière are listed overleaf:

---

* The method of lunar distances had to await solution by Nevil Maskelyne in 1763. (7)

**As late as 1714 the observed unreliability of a watch at sea was attributed to the following causes: the motion of the ship, variations of temperature, variations of humidity, variations of gravity at different latitudes.

London 6°39' error in longitude (easterly)

Cape Clear (S.W. Ireland) " " 4°30' (easterly)

Cape St. Vincent   "   " " 6° (westerly)

These last two points were landfalls for seamen and the errors in their position were significantly large.[9]

In principle, any fast-moving astronomical body can be used as a basis for time-keeping, provided its motion can be predicted and it can be accurately observed. Galileo used pendulum methods for the timing of astronomical observations, in particular the occultations of the satellites of Jupiter, and in 1616-1617 put forward this method for the determination of longitude in an effort to win the prize offered by King Philip III of Spain.

As soon as telescopic sights were applied to astronomical instruments, the resolving power of human observation was increased enormously. Galileo's object glass at Arcetri had a resolving power of between 3"-10" of arc. Astronomical events could subsequently be timed with greater accuracy. In 1616 and 1617, Galileo negotiated with Spain over the navigational uses of his astronomical discoveries. This had no immediate result but the longitude problem on land was now capable of solution. Galileo made a series of observations of Jupiter and its satellites throughout the planet's 12 year orbit. These ceased in 1619, but with the assistance of Father Renieri of Pisa University, his tables of remarkable accuracy were completed but never published.[10]

3.3 THE LONGITUDE PROBLEM AND THE ENGLISH SCENE IN THE SEVENTEENTH CENTURY

Partly because of the poor quality of maps and charts and the necessity of finding the longitude, Louis XIV in 1666 founded L'Académie Royale de France and the Paris Observatory was founded in 1667. Through the observational work of Jean Dominique Cassini and Jean Picard, it was possible to predict the times of the eclipses of Jupiter's satellites, so that comparison of the local times at which observations of the same event were made would give the longitude of the observer. From 1676 onwards this method was extensively used on land for determining and mapping the longitudes of many places in Europe, Africa, Asia and America. Picard issued the first edition of Connoissance des Temps for 1679 and predictions of the eclipses of Jupiter's satellites were included from 1690. The practical difficulties of observing the satellites of Jupiter from a moving ship, especially using the long telescopes that were then in favour were too great for the method to be of any use. During the early years of the Royal Society, the problems of navigation were very much to the fore. Among the experiments of 1662 were the trials of a pendulum clock which had been specially designed for finding the longitude at sea by Huygens in 1659. Henry Gellibrand (1597-1637) followed Edmund Gunter as Professor of Astronomy at Gresham College. He is best known for his observations establishing the secular variation of the compass needle.[11]

Henry Bond (1600-78), a teacher of navigation, began

his experimental work on magnetism after Gellibrand had
announced the variation of the variation in 1635. He was
editor of the Seaman's Kalendar for about 20 years and he
successfully predicted the annual rate of the variation
of the variation.* When a Committee for the Variation
was formed with Sir Robert Moray as Chairman, Henry Bond
was invited to membership.

In 1661, Thomas Streete (1621-89) re-edited his most
important work, Astronomia Carolina, in which he promised
a method for longitude based on perfecting his new lunar
tables. Lord Brouncker and Sir Robert Moray were asked to
examine Streete's book before a copy was presented to King
Charles II. During the year 1674, a young Frenchman at
the English Court, the Sieur de St. Pierre, announced that
he had a method for finding the longitude, and the Committee
for the Variation was asked to examine his claim. St. Pierre
was probably a protege of Louise de Kerouaille, Duchess of
Portsmouth, and he petitioned the King to have data supplied
for testing his method. St. Pierre's method, like
Streete's, was that of lunar distances or the moon's
appulses to the fixed stars. Meanwhile, Henry Bond had
completed elaborate tables of magnetic elements, including
dip, by which a magnetic network could be drawn over the
globe. The Committee advised the King to make a grant of
£30 to Bond to have dip-needles or 'inclinatories' made to
test his theory. At this time (1675), John Flamsteed,

---

* Bond's value for the variation in Whitehall on June 8th
1665 was 1°22'30" W and on June 1666, 1°35'36". On the
basis of these measurements he calculated the westward
revolution of the poles to have a period of 600 years.

who had made himself expert in this question of the motions
of the moon was then staying with Sir Jonas Moore at the
Tower of London.*   Flamsteed was co-opted to the Committee
and drafted a report explaining why St. Pierre's method was
unsound.   Within a week, Charles II, informed that even
the catalogue of fixed stars was of dubious accuracy,
authorised the foundation of the Royal Observatory.   On
June 30th, Moore and Hooke accompanied by the nineteen year
old Edmond Halley, went down to Greenwich Park and selected
a site on the green knoll overlooking the river.   The
warrant dated 22nd June 1675 states that the observatory is
to be built,

> "... in order to the finding out of the longi-
> tude of places and for perfecting navigation
> and astronomy."   (13)

John Flamsteed (1646-1719), a young Derbyshire clergyman, was appointed as the first Astronomer Royal.
Flamsteed's observations extending over 44 years were
published after his death as Historia Coelestis Britannica
(3 vols, 1725).   This work is the foundation of modern
fundamental or positional astronomy.

The work of Cassini and Picard at the Paris Observatory and its application to mapping gave rise to concern
about the inaccuracies of the available land maps of Britain.
In 1668, Jean Dominique Cassini published his Ephemerides
Bononiensis Mediceorum Siderum which was the result of

---

* Sir Jonas Moore was a member of the Committee for the
Variation and was Surveyor General of Ordnance since 1673.
He took part in the survey of London after the Great Fire
of 1666.   He was a governor of the mathematical school at
Christ's Hospital. (12)

sixteen years of observational work in Italy on Jupiter and its satellites. He had measured the time with a pendulum controlled clock and his tables gave their times of immersion and emersion in hours, minutes and seconds. In 1676, his enlarged _Ephemerides_ suggested a world survey using Jupiter's satellites to determine the longitude of places. This proposal appealed to Louis XIV and the task was promptly begun. Often dramatic changes to existing maps resulted.

As Marguet says:

"ces phenomènes, (the eclipses of Jupiter's satellites) - permirent de rectifier d'énormes erreurs, de l'ordre de 20°, sur la longitude de lieux très éloignés des observatoires europèens. La géographie fut renouvellée." (14)

It is said that Louis XIV was not wholly pleased by the results of these surveys as the Atlantic coast of France was thrust far to the east, apparently diminishing the extent of his domain. A map of France showing the improved outline of the coast after the survey of Cassini and Picard is in the National Maritime Museum, and shows clearly the reason for Louis' displeasure.(15)

Greenville Collins had begun a survey of the English and Scottish coasts in 1680 at the partial expense of the Crown; but by the Spring of 1683, Collins complained that the money promised by the King had not been forthcoming and the Admiralty showed no interest in the enterprise. In 1678, Robert Hooke wrote to Isaac Newton,

"I am informed likewise from Paris that they
are about another work, viz: of settling the
longitude and latitude of considerable place:
the former of those by the eclipses of the
satellites of Jupiter.   MM Picart and De La
Hire travel, and Msr. Cassini and Romer observe
at Paris ...   I have written to a correspondent
in Devonshire to see if we can do something
of the kind here."   (16)

Hooke's plans never came to fruition, but in 1680 John Adams
published his large maps of England and Wales with the
Index Villaris.   With the approval of the Royal Society,
Adams attempted a geodetic survey of England and Wales from
a measured base-line in Somerset.   Simultaneously a coastal
and topographical survey was to be undertaken and the
'angles of position' of some 40,000 prominent land-marks
to be measured.   Expenses far exceeded expectation and
subscribers did not come forward.   The political troubles
following the accession of James II put an end to the
enterprise.

Picart's Connoissance des Temps was the true fore-
runner of the Nautical Almanac and was published under royal
patent in 1687/9; it became immediately known and used in
England since the Royal Observatory and the Paris Academie
were in close correspondence.   The Connoissance des Temps
gave the longitudes of 86 towns, taking Paris as the Prime
Meridian.   The eclipses for the year were taken from
Kepler's Rudolphine Tables and were followed by a table of
the entry of the sun into the successive signs of the
zodiac, the true places of the planets at five-day intervals
and that of the moon for every day in the year.   The length
of a degree was given as 57,062 toises or 69.1175 miles.

The magnetic variation as found by De La Hire at the Paris Royal Observatory was also given. The publication was taken over by the Bureau des Longitudes in 1798, and has appeared annually ever since.

Edmond Halley had been commissioned as a Master and Commander in the Navy in 1698 and between 1698 and 1700 completed two Atlantic voyages to determine the variation of the compass. In 1701 he published a chart of magnetic variation in the Atlantic Ocean and in 1702 issued his first magnetic world chart. Halley's interest in magnetic phenomena probably originated during his three month voyage to St. Helena 1673-1674. He had noted that in latitude 15°N, the dipping needle lay horizontal. In June 1683, Halley contributed a paper to The Philosophical Transactions entitled A Theory of the Variation of the Magnetic Compass and in 1691-1692 in a series of papers to the same journal gave his four-pole hypothesis of terrestrial magnetism. Henry Bond had correctly calculated a zero variation in 1657. Bond postulated 600 years for the period of the westward revolution of the moveable poles and Halley later revised this to 700 years.[17] Halley's papers stimulated interest in the study of terrestrial magnetism and observations were forwarded to the Royal Society from all over the world.[18]

In 1691 several warships were lost off Plymouth, having mistaken the Dodman for Berry Head. Sir Cloudesley Shovel, returning from Gibraltar with his fleet in 1707, had cloudy weather during practically the whole passage, and after some twelve days at sea took the opinions of the navigators of all his ships as to his position. With one exception, their reckonings placed them in a safe position west of Ushant, and he accordingly stood on: but the same night, in fog, they ran on the Scillies. Four ships were lost, and nearly two thousand men, including the Admiral himself.[19] Disasters such as this brought to the public notice the urgency of the longitude problem. Numerous proposals were made offering solutions both traditional and novel. Appendix No. II to this chapter lists some of the printed tracts which appeared during the first half of the eighteenth century.

It is at this stage that Whiston and his colleague, Humphrey Ditton (1675-1715) enter the story. Humphrey Ditton was formerly Master of the New Mathematick School in Christ's Hospital, London.

3.4 THE ACT OF 1714 AND WHISTON'S PROPOSAL

In July 1713, Whiston and Ditton wrote to Joseph Addison (1672-1719) announcing their important discovery concerning longitude.

> "We are well satisfied that the discovery we have to make as to this matter is easily intelligible by all, and ready to be practised at sea as well as at land."

> "... We are ready to disclose it to the world if we may be assured that no other person shall be allowed to deprive us of those rewards which the publick shall think fit to bestow for such a discovery." (20)

At the end of April 1714, Whiston and Ditton petitioned the House of Commons that a Bill should be brought before

Parliament to appoint a suitable reward for whoever would
devise a method for the discovery of the longitude. This
petition was received on 29th April 1714.

> "The House appointed a Committee to consider what
> encouragement was fit to give such as should find out
> the Longitude; which Committee, having on 4th of
> June, asked Mr. Whiston and Mr. Ditton some questions,
> in the presence of Sir Isaac Newton, Dr. Halley and
> some other celebrated Mathematicians, came to these
> two Resolutions,
>
> 1. That it is the Opinion of this Committee that
>   a reward be settled by Parliament, upon such
>   person or persons as shall discover a more
>   certain and practicable method of ascertaining
>   the Longitude, than any yet in practice, and
>   that the said reward be proportioned to the
>   Degree of Exactness to which the said method
>   shall reach.
>
> 2. That the House be moved, that leave be given
>   for a Bill to be brought in accordingly."   (21)

On 11th June, a report from the Committee was read
to the House of Commons and the comments of Sir Isaac
Newton, Dr. Clark, Mr. Halley and Mr. Cotes on the proposals
of Whiston and Ditton give an informative picture of existing methods of determining longitude.

> "Sir Isaac Newton attending the Committee said
> that for determining the Longitude at Sea, there
> have been several projects true in theory, but
> difficult to execute:
>
> One is by a watch to keep time exactly: but
> by reason of the motion of a ship, the variation of
> heat and cold, wet and dry, and the difference of
> gravity in different latitudes, such a watch hath
> not yet been made:
>
> Another is by the eclipses of Jupiter's
> Satellites: but by reason of the length of the
> telescopes requisite to observe them, and the
> motion of the ship at sea, those eclipses cannot
> yet be there observed:
>
> A third is by the place of the moon: but her
> theory is not yet exact enough for this purpose:
> it is exact enough to determine her longitude
> within two or three degrees, but not within a
> degree.

A fourth is Mr. Ditton's project: and this
is rather for keeping an account of Longitude at
sea than for finding it if at any time it should
be lost, as may easily be in cloudy weather.

. . . .

In the fourth way, it is easier to enable seamen to know their distance and bearing from the
shore, 40, 60 or 80 miles off, than to cross the
seas; and some part of the reward may be given
when the first is performed on the coast of Great
Britain, for the safety of ships coming home: and
the rest when seamen shall be enabled to sail to
an assigned remote harbour without losing their
Longitude, if it may be." (22)

Whiston and Ditton circulated a broadsheet entitled

Reasons for a Bill Proposing A Reward for the Discovery of

the Longitude, dated 10th June 1714.

"June 10th, 1714.

REASONS for a Bill, Proposing a
Reward for the discovery of the
Longitude.

I.   This Bill is unexceptionable, because it is general,
and not confin'd to any one Project, Person, or Method;
but gives equal Hopes to all Judicious Proposers whatsoever.

II.   Because in this Bill no Money is insisted on before
any Method for the Discovery of the Longitude is, upon
Trial, actually found Practicable and Useful.

III.   Because Sir Isaac Newton's own Paper, delivered into
the Committee, gives hopes, that the Known Method by the
Theory of the Moon, which is hitherto not exact enough, may,
upon due Encouragement, in time be brought to Perfection.

IV.   Because the Method now propos'd is own'd by all, to
whom it has been communicated, to be certainly true in
Theory.   It cannot therefore be fit to have it conceal'd,
even tho' it were not yet known to be practicable; because
in that Case, future Improvements might still make it so.

V.   Because its great Use at Land and in Geography is
indisputable, and was distinctly observ'd by Sir Isaac
Newton and Dr. Halley, upon the first Proposal of this
Method to them.   And we beg leave to say, that this Use
alone is so great and extensive, that if there were no other,
it would highly deserve the Encouragement of the Publick.

VI.   Because another great Use is also undoubted, viz, for
all Places in the Narrow Seas, and within about 100 Miles of

all Shores and Islands; that is, for all Places where Ships are in the greatest Danger; as Sir Isaac Newton own'd to the Committee; so that if this Method extended no farther, yet it would highly deserve the publick Encouragement.

VII. Because there is little or no Reason to doubt of its Use at any Place at Sea, even where Ships are allowed to be in the least Danger; since in the most doubtful Case of All Sir Isaac Newton has, in his Paper delivered to the Committee, proposed a most effectual Remedy for the same; as will be clearly understood when the Method it self is known to the World.

VIII. Because this Method will save the Nation great Sums of Money, which the Want of it does now occasion; as will appear upon Trial.

IX.   Because the Charges of it will be inconsiderable, in Comparison of the Advantage; as will also fully appear upon Trial.

X.   Because it will prevent the Loss of abundance of Ships and Lives of Men; as it would certainly have sav'd all Sir Cloudesley Shovel's Fleet, had it been then put in Practice.

XI.   Because it is easy to be understood and practis'd by Ordinary Seamen, without the Necessity of any puzzling Calculations in Astronomy.

And we take leave to Recommend the Learned Savilian Professor of Geometry at Oxford, Dr. Halley, as the fittest Person in the World for the Tryal, and Practice, and Improvement of this Method; and do hereby Declare, that we are willing that he go equal Shares with us in the Reward, if he please to undertake so Useful a Work, and the Publick please to make that Reward equivalent to the great Dignity and Importance of the Discovery.

Will Whiston.
Humphrey Ditton.

June 10 1714. "   (23)

In 1714 the Government passed an Act (13 Ann. cap.14), offering a graduated scale of rewards for any "generally practicable and useful" method of finding longitude at sea. The criterion of the method was to be the amount of error in determining a ship's longitude at the end of a six weeks' voyage. If this error did not exceed sixty geographical miles, the inventor's reward would be £10,000; if less than forty miles, £15,000; if less than 30 miles, £20,000.[24]

The Board of Longitude was a permanent body of 17 Commissioners set up by the Government in 1714 and charged with supervising all competitions for the rewards and empowered to advance small sums in support of methods which appeared promising. They met about three times a year at <ins>after 1760</ins> the Admiralty. The Board was finally dissolved in 1828. In all, they had disbursed a large sum of money, about £101,000.[25]

Most printed submissions to the Board of Longitude were preceded by a list of the contemporary Commissioners.

Whiston and Ditton presented their proposal for the determination of longitude by a method of simultaneous flashes and sounds from stationary ships, <u>A New Method for Discovering the Longitude both at Sea and Land</u> (1714).

> "It was proposed to fix stationary ships or buoys
> at least at the distance of 600 geographical miles
> or ten degrees in all parts of the ocean.   In these
> ships, a mortar or great gun was to be exactly fired
> every midnight, which being heard by the navigating
> ships, the mariners are supposed to know their dis-
> tance by the degree of the sound, or for their
> further assurance bombs were to be thrown up as
> high as possible, the utmost altitude of which
> being seen by the fire and observed by the mariners,
> their distance is found by the difference of the
> altitude from the known one of the stationary
> ship;   or still further, by firing a gun at the
> moment the bomb arrives at its greatest altitude,
> the same distance will be found by observing the
> difference of the time between hearing the first
> and that of seeing the last;   the same may be done
> if the sound and light are made at any given
> interval."   (26)

In other words, Whiston and Ditton proposed that permanent floating light-ships should be established on the principal trade routes, firing at midnight star-shells arranged to explode at a height of 6,400 ft., thus affording ships an opportunity of determining their distance from the nearest light ship by timing the interval between the flash and the report. They added that this method would be of particular use in the North Atlantic and on the authority of the Earl of Stanhope stressed that the North Atlantic did not exceed 300 fathoms in depth.*(27)

Newton had hinted to Whiston that his sound and light method might be of use in finding the longitude on land and in making exact surveys and maps of countries. In 1714, Whiston had the idea of making such a survey of England and he proposed to improve William Derham's table of the velocities of sound. He used the rectilineal canal, the New Bedford river, in the Isle of Ely and the old Roman road, Watling Street, which passes in almost a straight line through the flat part of Stafford, for this purpose.(28)

Whiston corresponded with Roger Cotes, Plumian Professor of Astronomy at Oxford, and a member of the Board of Longitude, about his trials with simultaneous flashes and sounds. He hoped to get persons in every market town to make the necessary observations.

Cotes replied to Whiston on 2nd December 1714,

"I have received your proposals and have disposed of them according to your desire. I have hitherto sent you no account of your light because I could not yet see it. I fear the distance of 60 miles is too great for your mortars. It will be very difficult for you to find persons in every market town capable of making the requisite observations ... Unless they be precisely and punctually instructed in these things and be furnished with such

---

* In fact, the average depth of the North Atlantic is about 2,000 fathoms and the maximum depth is 3,450 fathoms.

instruments as you think proper and moreover
be faithful and diligent in the work: your map
must needs be a lame performance." (29)

In April 1715 Whiston again writes to Cotes about the con-

tinuation of his experimental work on a survey of England:

"I send you the new edition of one proposal about
the Longitude and our survey of England ....
I think to move ere long for £500 (from the Board
of Longitude) towards past charges, the present
survey and some tryals at and perhaps near the
coast: and I hope I shall not be denyed." (30)

In January 1715, Whiston wrote to Samuel Barker, his future

son-in-law at Lyndon, near Oakham, Rutland:

"London Jan 27 1715

"The Lord Chief Justice King kept my paper about
a month without reading it: tho' I gave him
some account of it at last. I meet with few
or no objections against our scheme, in the matter.
I intend to move shortly for publick money to
plant a Gun at the Land's End for the Longitude
near the shores; in order to claim the one half
of our reward: reserving the rest of the claim
to hereafter." (31)

Whiston's 1714 proposal was almost certainly inspired by

William Derham's (1657-1735) experiments on the velocity

of sound. Derham carried on a correspondence with John

Flamsteed relating to experimental work on the velocity of

sound, and wrote in 1707,

"I am very desirous to perfect all my enquiries
relating to sounds, thinking it matter not as yet
sufficiently cleared either by others or myself.
And I believe I should absolutely master it could
I procure a gun to be fired for me during evenings
at Woolwich, either at the Church or on the Wharfe,
where the mortars stand. You having acquaintance
and interest among the Office of Ordnance or among
others who would do the thing I wish you would
procure this favour for me and for the promotion
of natural knowledge. If you can effect it, be
pleased to send me a word of it and I will give
special directions to the gunner." (32)

Derham had written for some exact information from
Flamsteed,

"What was the exact distance you measured and
what the exact number of seconds that the sound
spent in flying that distance, when you, (and as
I remember), Mr. Halley, tryed the experiment of
the Flight of Sounds ... I desire you to tell
me how the wind favoured or opposed the sound.
I find my experiment and yours greatly accord by
what I do remember you told me... I have a great
mind after Harvest to get Mr. Hudson along with me
to Foulness Sands to try the Flight of Sounds
with the greatest accuracy."   (33)

During the year 1714, Abraham Sharp (1651-1742)
corresponded with Flamsteed, whom he had served as an
assistant between 1676 and 1690.   Sharp refers to some
pamphlets and papers about Longitude which Flamsteed had
sent to him.

"Horton 1714

"... there is little or nothing tending to instruction or the advancement of science in any of them
except Mr. Whiston's, though I very much fear his
project being so changeable and impracticable at
Sea will scarce turn to an advantageous account
to himself. The Newspapers gave notice of his
proposing to make an experiment by causing a Ball
of fire to be shot up into the air every Saturday
evening at eight o'clock upon Black Heath which
being so near you cannot but see or hear it,
though you are too near and we are too far to make
any advantage of it yet would gladly know if
possible whether any others at a competent distance may have observed. I presume, if they
have it will certainly come to your knowledge." (34)

On April 3rd 1714, Sharp again wrote to Flamsteed
about Whiston's proposal.

"... tis probable they meet not with the encouragement they expected else methinks something
further would ere this have been procured."   (35)

After the death of his colleague Ditton in 1715,
Whiston continued his experiments with gunshots and flares

and issued a broadsheet announcing the firing of 'balls of Light' on 15th and 17th Days of May 1715.

"Broadsheet on Flares"

Printed.   Mr. Whiston hereby gives notice, that on Wednesday and Friday nights next, being 15th and 17th Days of this present Month of May, exactly at half an hour after ten, and at eleven o'clock, there will be two balls of light or fire thrown up from a Mortar at Hamstead Heath.   And he earnestly desires that all curious persons, within sixty miles distance, will please to observe the same, and to send him or some friend of his, word the next Posts, how plainly, and how many seconds the balls are seen;   and if they well can, at what utmost Altitude, and in what Angle from the Meridian, they are seen: and in places within twenty or thirty miles, how many seconds the sound is heard after the first sight of the Light also.   But Note, that if either of those nights prove misty or rainy, the projections will be deferred till the nights immediately following. A smaller ball of light or fire will each night be thrown up also, from a lesser mortar, at three quarters past ten."   (36)

Whiston sent a copy of this broadsheet to Flamsteed with a personal request that his experiments on Hampstead Heath should be observed at the Royal Greenwich Observatory.

Crosthwaite, assistant to Flamsteed made records of the experiments.

"In MS.

At the observatory, May 15th, the weather being somewhat rainy but calm and the wind westerly, at 10.30 o'clock we see the flash and heard the sound (but could not see the ball which was 47" in coming.)
May 17th misty.
May 18th.   The weather being clear and calm, at half an hour past ten we saw the flash and heard the sound, but could not see the ball [fireball].   At 10.30h we see the flash and heard the sound and see the ball of light for about 5" or 6".   The sound was very low and could not be heard above 2 miles further.   At 11 h. we saw the flash and heard the sound, but could not see the ball in the Air.

June 5th.   At 10^h 15', the weather being clear
and calm, we see the first ball, but could not see
the flash or hear the sound.   At 10h 30" we saw
the ball in the Air for about 15" or 16" but could
not see the flash nor hear the sound.   At 10h 45',
we saw the ball but could not hear the sound nor
see the flash.   At 11h 00, we saw the ball in
the Air for about 24" or 25"; but could not hear
the sound nor see the flash."

N.B.   The first two nights were with a mortar of 7 or 8
inches the last were with a mortar of about 4 inches. (37)

In 1714 (date not clear) George Young wrote to
Flamsteed saying that he had seen Whiston's book on Longi-
tude and understood that Flamsteed and the rest of the
Trustees judged it impracticable. (38)

A second edition of A New Method for Discovering the
Longitude both at Sea and Land (1715) was preceded by a
historical preface in which Whiston traces the development
of his ideas.

"Above a year ago, Mr. Whiston and Mr. Ditton,
with some other common friends spent part of an
afternoon and an evening together.   Mr. Ditton
took an occasion among other common discourse,
to observe to Mr. Whiston, that the nature of
sounds would afford a true method for the discovery of the Longitude, since the difference
between the apparent time where the sound is
made, and where it is heard; abating only the
time for diffusion, which was now well known;
is the difference of Longitude of those two places
in time."   (39)

Whiston goes on to say that he remembers having
plainly heard the firing of great guns above 90 or 100 miles,
when the French fleet was engaged with ours off Beachy Head
in Sussex (AD 1690) when he was in Cambridge.   On 7th July
1713, Whiston was present at a great Firework Display,
a Thanksgiving celebration for the Peace of Utrecht.   This
event seems to have inspired him to combine flashes of light

at great elevations with simultaneous reports from the
mortars so that visual and aural signals combined could be
detected by persons on land or at sea. This method was
to be applied to improve the state of cartography.

> "... He (Ditton) did also first observe that
> great use of our method at Land, in the Survey-
> ing of Countries, for the Perfection of
> Geography; which was also readily taken notice
> of by Sir Isaac Newton and afterwards by Dr.
> Halley, and that both of their own accord, upon
> our first communication of our method to them." (40)

Whiston calculates the possibility of extending his
sound and light method for distances of up to 200 geographical miles by placing the guns on elevations and firing
his 'balls of fire' to great heights. Tables correlate the
distance of a ship from the light signal by a measurement
of the angle above the horizon at which the ball of light
is visible. Two such tables are given:

(1) for explosions at 6,400 feet and

(2) for explosions at 10,000 feet.

A table of distances from the source of the sound correlated
with the time interval between the observation of the flash
and the arrival of the sound. Whiston takes this table
from William Derham's table of experimental data.[41]

The method of calculating longitude by this sound and
light method is demonstrated by worked examples. Whiston
closes his treatise with a plea for agreement about the
prime Meridian,*

---

* The Greenwich meridian was not agreed on internation-
ally until 1884.

"It is farther humbly proposed to the learned,
whether it may not be proper for all nations upon
this occasion, to agree upon one first meridian
or beginning of Longitude, for the common benefit
of Geography? and whether it may not be proper,
in that case, to fix it to the Pike of Tenariff,
as the most noted place already; and as the
place where the highest and most generally useful explosion by this method be made every midnight for the Discovery of the Longitude itself."[42]

Whiston's long continued interest in the longitude
problem was always accompanied by a consciousness of the
urgency of the preparation of improved maps, particularly
of the coast line.*

Derham had written to Flamsteed in 1703,

"... Mr. Townley's letter was chiefly about some
errors in Rain Tables he formerly sent me, but
there was one thing worth knowing if you have
not already heard it, viz, that Mr. Adams (John)
told him twas his opinion that Saxton's Mapps
(from when Towneley thinks Speede's were drawn)
have not misplaced any place in England above a
mile either in longitude or latitude." (44)

No major survey had been undertaken in England since
the late sixteenth century mainly due to lack of sufficient
funds but also because of the uncertainty in known methods
for determination of longitude. In 1715, Whiston had
published proposals for a survey of England and Wales as
advertised at the end of the second edition of his 1714
Proposal on Longitude.

---

*   Christopher Saxton's maps, a general map of England and
     Wales and thirty-four county maps appeared between 1574
     and 1579.   Saxton's remarkable rate of production leads
     one to conclude that he must have relied on earlier
     topographical works.   (43)

ADVERTISEMENT

Lately published, Proposals for a New Survey
of England and Wales according to this Method for
Discovering Longitude; and for a new sett of
Correct maps of the several counties, according
to such a survey, by Subscription, at Two Guineas
the whole sett, bound in Pastboard, are to be had
gratis at Mr. Whiston's in Cross-Street, Hatton-
Garden; at Mrs. Ditton's in Christ's Hospital;
at Mr. Senex's at the Globe near Salisbury Court,
Fleet Street; at Mr. Hauksbee's in Crane Court
near Fetter-Lane, Fleet Street; and at the booksellers of London and Westminster and of several
county towns.

These proposals were never implemented due to lack

of available funds and also because Whiston's method for

determination of longitude was recognised as inefficient

and impracticable.

3.5   EARLY WORK WITH DIPPING NEEDLES

Stimulated by the publication in 1581 of Robert

Norman's The Newe Attractive and William Borough's A Dis-

course of the Variation of the Cumpas, William Gilbert of

Norwich began in 1582 to study magnetism and the nature and

properties of the lodestone. In 1600 he published his

De Magnete. Although the major part of De Magnete was

devoted to the problem of the nature of electricity and

magnetism, Gilbert also dealt with the practical application

of the latter to the problem of navigation. Gilbert's

theory of variation was of interest. He rejected the

current theory that the magnetic poles were situated some

20° distant from the geographical poles, at determinable

points, and that this was the cause of variation. Instead

he thought that variation was caused by the proximity of

land masses, and by the varying depth of the ocean. While
this idea cleverly accounted for the now generally acknowledged irregularity of the distribution of variation, which
he held to be unchangeable at any place, it did not seem
promising to those who hoped to use the asymmetric theory
of the magnetic poles as an aid to finding longitude. But
in magnetic dip, he saw a possible means of determining
latitude independently of dead reckoning and celestial
observations.[45]

Norman had devised dip-needle mounted on a pedestal
(see below).

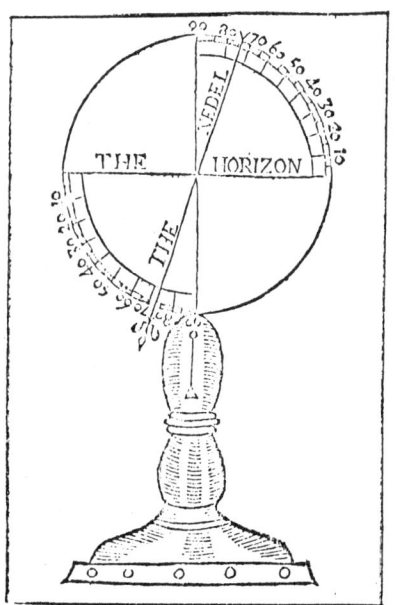

FIGURE 3

Norman's pedestal
Dip-needle as in
The Newe Attractive
(1581).*

---

* Norman gave 71°50' as the angle of Dip for London in
1576 and 11°15'E as the variation in the same year.

William Barlow modified this instrument for use at sea by fitting it with a suspensory thumb ring and by enclosing the sides of the dip-needle case, which measured about 5½" in diameter and 1½" in breadth with discs of unbacked Venetian mirror glass.*

> "I was the first that made the inclinatory instrument transparent, to be used pendant, with a glass on both sides, and a ring on the top ... and moreover hanged him in a compass box, where with two ounces weight, he will sit for use at sea." **

Suspended in the cock-pit where it was out of the wind, and aligned north and south, the dip needle could be read quite accurately.

Gilbert's contribution was to add an horizon line of brass and to incorporate the description of the instrument into his work. He also invented a second instrument to be used in conjunction with the dip needle. This was a disc of brass, or of pasteboard, on which were engraved "spiral" dip lines and parallels of latitude. By means of this instrument and a ruler, the latitude of a place could be determined mechanically (it was hoped) by means of its dip.[47]

---

* William Barlow (1544-1625) carried out experiments on the lodestone from 1579 onwards and interested himself in all matters relating to navigation. [46]

** William Barlow: <u>Magneticall Advertisements</u>.

FIGURE 4

Gilbert's Standing Dip-Needle.   From De Magnete,(1600)

The dip-dial was not easy to construct.   Edward
Wright, an enthusiastic improver of navigational aids and
methods, particularly of magnetic ones, was anxious to get
the Gilbert-Barlow dip instruments before the seafaring
public.   Thomas Blundeville, a friend of Gilbert, published
The Seaman's Kalendar (1602) with an appendix entitled
The Making, Description and Use of Two Most Ingenious and
Necessary Instruments for Seamen, for finding out the
latitude of any place without the help of the Sunne, Moone
or Starre.   The most valuable part of this appendix was
the final sheet.   This embodied the results of Henry
Briggs' (Gresham Professor of Geometry) laborious calculations, doing away with the need for a dip-dial.   Four
columns of figures tabulated the angle of dip calculated
for every degree of latitude between 1° and 90°.   Thus the
co-ordinated efforts of four men of science, (Norman,
Barlow, Gilbert, Briggs) resulted in their devising, out of
a complexity of research and calculation, a useful aid for
seamen.   One of the earliest surviving dipping needles is
a 12" Dip Ring from the seventeenth century now in the
Department of Natural Philosophy at the University of
St. Andrews.[48]

In 1608, Hudson took a good Inclinatory Needle with
him on his journey to the North Cape.   Hudson made several
highly valuable observations of the angle of Inclination,
from the latitude of 61°11' N to 75°22' N.   The values of
the Inclination ranged from 79° to $89\frac{1}{2}$° and so Hudson almost
reached the North Magnetic Pole.[49]

Henry Bond, already mentioned as a member of the

THE
LONGITUDE
AND
LATITUDE

Found by the

INCLINATORY
OR

Dipping Needle;

Wherein the
Laws of Magnetiſm are alſo diſcover'd.

To which is prefix'd,

An *Hiſtorical Preface*; and to which is ſubjoin'd,
Mr. ROBERT NORMAN's *New Attractive*, or
Account of the firſt *Invention* of the *Dipping Needle*.

By WILL. WHISTON, M. A.
Sometime Profeſſor of the Mathematicks in the
Univerſity of CAMBRIDGE.

Καὶ πᾶσαι πεδίοιο τείϬοι, κỳ πεύχεες ὄχϑαι,
Ὄυρεα ϑ' ὑλήεντα, κỳ ἄϧϧια κύματα πόντυ
ἜυϬατα δὴ, κỳ ἔυπλοα ἔϗεϑ ἤμαϑι κείνοις.
Orac. Sibyll. III. v. 715, 716, 717.

LONDON:
Printed for J. SENEX, at the *Globe* in *Saliſbury-Court*;
and W. TAYLOR, at the *Ship* in *Pater-Noſter-Row*.
MDCCXXI.

Committee for the Variation published a broadsheet in 1673 advertising his claims to knowledge of terrestrial magnetism.[50]

In 1673, John Pell (1611-1685) made a summary of relevant events in the study of terrestrial magnetism since the days of Edmund Gunter (see Appendix No. I to this chapter).

3.6   WHISTON'S   WORK   WITH   DIPPING   NEEDLES

Whiston's first printed work on the use of the Dipping Needle for the determination of Longitude was The Longitude and Latitude found by the Inclinatory or Dipping Needle (London, 1719).   A second edition of this work in 1721 had an historical preface, and as an appendix Whiston added the complete text of Robert Norman's The Newe Attractive (see plate opposite).

Whiston was familiar with the work of Colonel Windham who made use of the dipping needle by sea and land about 1672.[51]   Whiston acknowledges that he made use of Windham's observations to determine the dip at Salisbury for use on his chart of isogonic lines.   Nicholas Fatio (1664-1753) (sometimes spelt Faccio) assured Whiston that Boyle had used a terrella and a small inclinatory needle.

Whiston deplores the fact that Halley did not carry a Dipping needle on his Atlantic journeys 1700-1702.   In 1700 the Reverend James Pound went to Madras in the service of

---

* Dr. John Pell corresponded with Gellibrand and Mersenne on the secular variation of the magnetic compass.

the East India Company as chaplain to the merchants of Fort St. George, accompanied by James Cunningham. According to Whiston they went to China and for a great part of the journey made observations with a dipping needle.[52] In 1706, Père Noel S.J. sailed from Lisbon to the East Indies as a missionary and all the way made accurate observations with a good inclinatory needle. These observations were published at Prague and Whiston believes them to be the best readings available. Further observations were available from the voyage of Louis Feuillee (1600-1732) who in 1708 observed the inclination with great care and exactness from 13°S to 36°S and recorded values of Dip ranging from 18° to 55°. Samuel Molyneux (1687-1728) passed on Feuillée's observations to Whiston. In November 1718 Whiston was visited by Christoph Eberhard from Saxony, pastor of Altona near Hamburg, who claimed to have a method for solving the Longitude by the Dipping Needle. Whiston found that Eberhard's ideas were devoid of originality and were based on Pere Noel's book. This hypothesis supposes two internal loadstones in the earth and by the insertion of small lodestones under maps or within terrestrial globes, the phenomenon of dip could supposedly be investigated.

Whiston states clearly that as late as November 1718 he had never seen a dipping needle.

> "I did not then distinctly know what a dipping needle was, nor do I remember that before that time, I had ever seen such an instrument in my life." (53)

Whiston judged that Mr. Eberhard's needles were so small as to be useless. Colonel Windham lent Whiston a

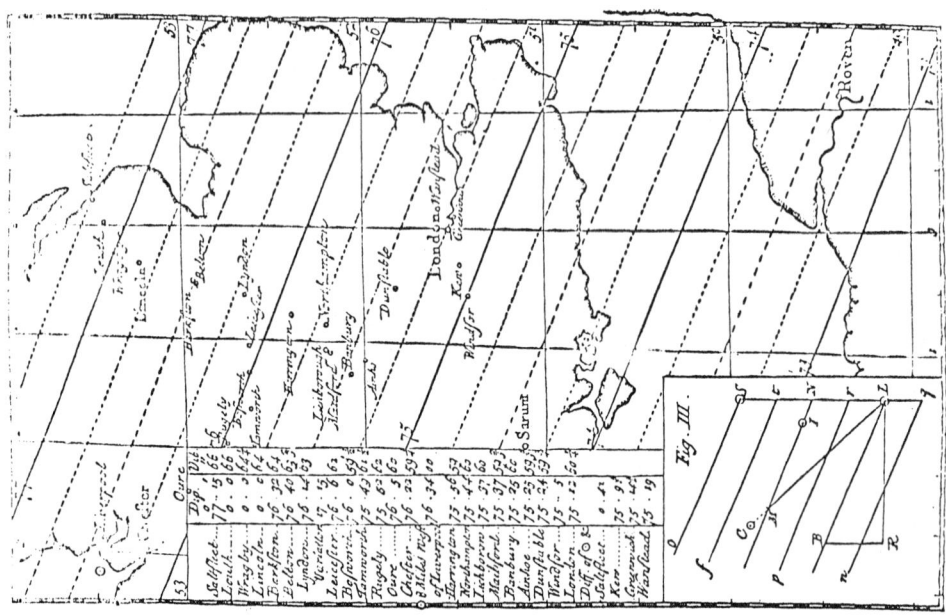

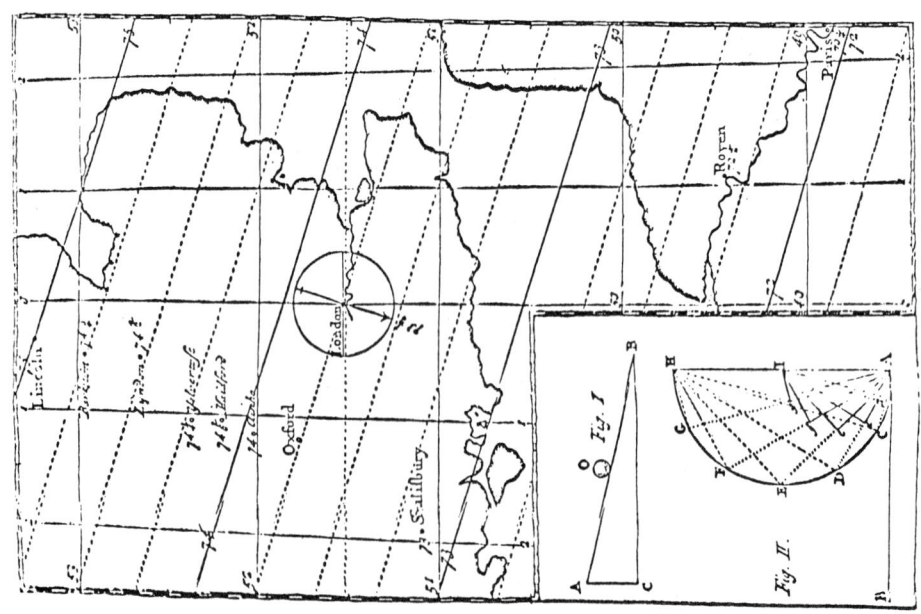

small dipping needle and trained him in the use of it.
Later Windham lent Whiston a larger dipping needle and with
this he first carried out his experiments.

After considering Halley's map of Variation and the
usefulness of this map to mariners, Whiston proposed the
compilation of a similar chart showing lines of equal
magnetic dip.

> "when therefore I considered, that the lines of
> equal Dip could hardly be more irregular than
> those of the variation; and well knew that
> their mutation was a great deal slower; and
> that these might probably be useful all over
> the world; I conceived great hopes, that this
> way of application of the power before us might
> very probably discover the longitude." (54)

The two charts in Whiston's 1719 work are thought to be the
oldest showing isogonic lines.[55] (shown opposite).

The observations for the first and earlier chart were
made using a 12" dipping needle while for the second chart
he used a dipping needle $47\frac{1}{2}$" long, probably the same needle
which he used for his demonstrations to the Royal Society.
G. Hellman also notes that Whiston refers to lines of equal
dip for the whole earth drawn on a globe. Whiston himself
refers to these lines of equal dip which he drew upon
"Mr. Molyneux's Terrestrial Globe".[56] We have only
Whiston's word for this statement as no trace of the globe
has yet been found.

It was to Samuel Molyneux* that Whiston entrusted the
journals of his own experiments for finding longitude at

* Samuel Molyneux (1687-1728) was secretary to The Prince
of Wales. He had a private observatory at Kew and was
one of Whiston's faithful patrons.

sea and it was in Molyneux's ships that the observations
were made.[57]   No further attempt was made to record lines
of equal dip over the whole earth until that of Johann Carl
Wilcke in 1768.[58]

As soon as Whiston began to pursue this use of the
Dipping Needle, he got in touch with Flamsteed, now an old
man of 72 years.

> "November 15th 1718
> "If we make and use such a map for the Inclination
> of the Dipping Needle as Dr. Halley has made for
> the variation of the Horizontal Needle, we obtain
> the Longitude.
> Will Whiston"
> "Will you please note the day when you receive this."[59]

The letter reached the Old Royal Observatory on the
same day and was witnessed by Abraham Ryley and Joseph
Crosthwaite, assistants to Flamsteed.   Flamsteed in his
failing hand made the following note on the outside fold of
Whiston's letter.

> "About a Map to show the Inclination of the Magnetic
> Needle.   Who put him on it?   Mr. Perkins (?)
> busy about it 1680.   Mr. Halley bought his
> papers (?)."

Whiston was not a Fellow of the Royal Society but
this did not prevent him from presenting his experimental
and theoretical work to them.   On June 23rd 1720, he was
admitted to the meeting of the Royal Society and in the
presence of Newton as President demonstrated some of his
magnetic experiments on which he based his proposed method
of finding the longitude by sea and land.   He made the
following points:

"1. That the earth has only two principal magnetical poles, one in the northern and the other in the southern hemisphere.

2. That these two poles are not diametrically opposite and each of them are at a considerable distance from the poles of the world.

3. That a magnetical needle situated in the plane of the meridian and librating upon an axis parallel to the horizon will constantly dip in the same angle of inclination to the horizon, in all such places as are situated under the same circle described about the nearest magnetic pole - excepting some inequalities which may arise from the unequal distribution of the magnetic matter and which are to be collected by experiments.

4. That the different angles of the dip in passing from the Magnetic Equator to the Pole observe the proportion of the sine of the distance from the Equator."   (60)

Whiston went on to suggest that maps be made with the lines of equal dip shown.   From these maps, it would be possible to deduce the longitude for any place in the world whose latitude was known.

"That the angle of dip may be observed to great exactness he proposes to have a needle of great length 6-8 feet and if the observation is to be made at sea, he proposes to place the instrument in the centre of the ship's motion, that is at the bottom of the mast where it would cut the plane of the water, this being the place where the motion of the ship least affects all bodies within it."   (61)

He went on to propose that a more accurate measurement of the angle of Dip could be made by comparing the periodic time of the vibrations of the dipping needle with the periodic time of the same needle vibrating in a horizontal position.

$$\frac{\text{Cosine Angle Dip}}{\text{Radius of Needle}} = \frac{\text{square of Periodic Time (Dip Position)}}{\text{square of Periodic Time (Horiz. Position)}}$$

Whiston demonstrated this with a needle four feet in length with the following results:

Periodic Time (Horizontal Position)   47 seconds

Periodic Time (Dip Position)   24½ seconds.

The Royal Society did not pass any verdict on Whiston's
proposals but the journal book records,

> "Mr. Whiston having advanced something which seems
> to require further confirmation by experiment,
> the Society thought fit to suspend their
> judgement."   (62)

It seems likely that Halley advised Whiston to carry
out further measurements on the angle of Dip at points more
widely separated over the British Isles.   Four months
later, Whiston wrote to Halley reporting on his measurements
of the angle of Dip at Saltfleet and Chester.

> "A Letter from Mr. Whiston to Docr. Halley about
> the Dipping Needles.   Dat'd - Lyndon Oct. the
> 14th 1720.
>
> Sr.
>   I have now according to your Desire and
> Recommendation, when I was at the Royal Society,
> tried my Dipping Needle both at the East and West
> Seas of this Nation, and that nearly in the same
> Parallel;   I mean at Saltfleet and at Chester.
> I have also tryed it for above 30 Miles in the
> Parallel of this Place, I mean at Tamworth,
> Bosworth, Leicester and Lyndon, to say nothing of
> several other intermediate Observations in my other
> Journeys;   I also took a friend with me all along,
> who was neither unskilfull in such Experiments nor
> Incurious about the Exactness of them, we stil
> found them to conspire, that in two Degrees of a
> great Circle the Dip: in England alters even in
> the same Parallel full 41!   Nor is there Generally
> any Room where the Experiment is nicely made for a
> mistake of 4, I have not neglected where ever I
> could, to Perform the Tryalls before the best
> Judges;   among whom was your Learned Friend Mr. Ward
> at Chester, who if you desire him will no Doubt,
> give you an account of our tryal there.
>   I beg the favour of the Comunication of this
> Account, to the President and the Rest of the
> Members of the Royal Society at their next meeting;
> who must be well pleased with the success of a

Discovery that lends so much to the benefit of
the Publick; it Plainly giving the Longitude
to 10 Miles.

I am

Will: Whiston.

Dip at Saltfleet	....	77°15'
Dip at Chester	....	76°22'
Difference	....	00°53'

Therefore the Dip about 8 Miles West of
Liverpool must be   ... 76°34'

At Saltfleet two Degrees distant
in the same Parallel   ... 77°15'

The Difference is two Degrees 00 41." (63)

At the meeting of the Royal Society on October 27th
1720 Whiston's letter was read.   The Journal Book notes
Whiston's claim to be able to determine the longitude
within 10 miles.(64)

Again on March 22nd 1722, Whiston demonstrated some
of his magnetic experiments to the Royal Society.

"He produces a small terrella with some fine needles.
By applying the needle to several parts of the
Terella, he offered to explain how the magnetical
needle dips in several parts of the earth and
endeavoured to show that the angle at which the
needle dips upon the surface of the Terella is
always proportional to the sine of the distance
from the Pole."   (65)

Whiston aroused great interest in the possibilities
of his dipping needle method for determining longitude.
A subscription list of 1721 indicated a total of
£470. 3s. 6d.(66)

In his memoirs, Whiston refers to the success which
followed his publication on the use of the Dip Needle.

"After the publication of this treatise, I found
so much encouragement from many benefactors that
I was enabled to procure some new observations
of the angle of Dip in several parts of the

World, in order to perfect this discovery;
the substance of which is printed at the end
of my <u>Calculation of Eclipses, without Parallaxes</u>.
Which upon the whole cost me a very great deal of
pains, to contrive the instruments, and hang them
in ships, so as to take the Dip, with an exactness
sufficient for my purpose." (67)

In Whiston's 1724 publication, <u>The Calculation of Solar</u>
<u>Eclipses without Parallaxes</u>, he adds at the end <u>An Account</u>
<u>of some late observations made with Dipping Needles, in</u>
<u>order to discover the Longitude and Latitude at Sea</u>.

Whiston's dipping needles were carried in four ships
undertaking voyages in 1722 and 1723. Captain James Jolly
travelling to Archangel, Russia; Captain Othniel Beal to
Boston in New England and later to Barbados, and Charles Town
in South Carolina; Captain Tempest travelling to Antigua
(West Indies) and St. Christophers (now known as St. Kitts);
at a later date (long after the rest), Captain Michel
travelling to Hamburg. Whiston goes on to list the conclusions which he has come to from the analysis of observations
made on these voyages.

"There is one spherical loadstone and but one
centre of our earth; and that this loadstone,
like any other spherical loadstone had but one
Northern pole; contrary to Dr. Halley's
Hypothesis." (68)

Whiston specifies the position of the Northern Pole.

"I take the northern pole of the Terrestrial
Magnet to be about the meridian of Archangel
in the latitude of 75½°."

In 1970, the magnetic pole was in latitude 81°56'.
There is evidence that in 1720 it was in latitude 87°, about
11° further north than Whiston's calculated location.

Whiston expresses his intention to determine by experiment
the position of the magnetic north pole.

> "Nor indeed shall I be at rest, till I have sent
> a dipping needle to Hudson's Bay, on purpose to
> determine this dispute (with Halley) about the
> four poles."   (69)

Whiston did in fact send a dipping needle to Hudson's
Bay on board ship in 1723.

> "Yet hearing nothing of it till the year 1739, I
> concluded the ship, in which it was, to have been
> long ago cast away and the Needle to have been
> long ago at the bottom of the Ocean.   But the
> last year the curious Captain Middleton, who was
> then the mate of that ship in which my dipping
> needle was sent;   assured me the ship was safe;
> and that he had himself tried the experiment 79°W
> of London at Cape Diggo in Hudson's Straits;   and
> found the needle stood strongly perpendicular at
> that place."   (70)

Whiston concludes his report by printing a table of
the angle of dip for every degree distant from the Magnetic
Pole and the Magnetic equator.   It appears that frames in
which these Dipping Needles were hung and which were designed
by Whiston to counteract the oscillations of the ship,
did in fact cause a great deal of trouble.

> "(Captain Michel) ... was not willing to encumber
> himself with (the Frame); and I suspected that
> in its present contrivance it did more than help
> the nicety of the experiments."   (71)

Between March 29 and May 2nd, 1723, George Graham
kept detailed records in the variation of the dip

> "I made likewise some experiments with the dipping
> needles to try if the dip and vibrations were
> constant and regular.   The needle I made for
> this purpose was 12 inches and one tenth long,
> half an inch broad in the middle, but not above
> one tenth near the ends."

Extensive carefully executed observations follow of both
the angle of dip and the time of vibration of the needle.
A study of the observations reveal that,

> "there is a very considerable difference both in
> the quantity of the dip and in the quickness of
> the vibrations." (72)

It was this experimental work of Graham which determined
Whiston to discontinue his own work with dipping needles as
a method for finding longitude.

A careful study of the 1721 edition of Whiston's
The Longitude and Latitude Found by the Dipping Needle leads
me to the conclusion that his careful experimental work must
have helped towards the compilation of more accurate, current
data about the magnetic declination and dip and at the same
time contributed to an improvement in design of dipping
needles.

The next major contribution to the investigation of
magnetic declination took place fifty years later, when
Edward Nairne made carefully constructed dipping needles for
the Navy Board to be carried by Captain Phipps in his
voyage towards the north pole 1774.

Whiston's theories, experiments and observations were
not carried out in isolation. Francis Hauksbee Jnr.
(1687-1763) who had assisted Whiston in his lecture course
of 1714, was between 1719 and 1726 working with John Hadley
on the improvement of reflecting telescopes.* Whiston must
have engaged the services of Hauksbee to help him with his
practical work on magnetism.

x see page 4.27

"... the needles having been weighed both before
and after (being touched with the loadstone)
with the same weights by our very accurate and
skilful operator Mr. Hauksbee." (73)

Whiston makes frequent reference to Lord Paisley,
author of <u>Calculations and Tables relating to the Attractive
Power of Loadstones</u> (1729) whom he refers to as "the great
reviver and encourager of the knowledge of magnetism among
us".   Lord Paisley gave Whiston the use of his two famous
terrellae, one of 2.8" diameter and a more powerful loadstone of 1.25" diameter.

The work of Dr. Brook Taylor on investigations con-
cerning <u>The Law of Magnetical Attraction</u>(74) was known to
Whiston as was also the experimental work of Rev. Benjamin
Worster (joint principal with Thomas Watt of Watt's
Academy, Little Tower St.) on

"finding the Law of attraction in Magnets ...
by comparing vibrations and oscillations of
needles with oscillations of a prismatic
pendulum of the same radius." (75)

Whiston appeals to Newton's comments on the Laws of
Magnetism (<u>Principia</u> Book I Sect. 13 Prop. 85 and Book III,
Prop 6, Cor. 5) that the attractive power must decrease
in more than a duplicate (i.e. inverse square) proportion
of the distance.   Whiston's own experimental work led him
to believe that magnetic attraction decreased according to
the sesquiduplicate power of the distance.

Force of attraction (or repulsion) is proportional to $\dfrac{1}{(\text{distance})^{5/2}}$

When referring to the nature of magnetism, Whiston favours
the notion of a magnetic fluid

"circulating continually around the loadstone"[76]

but, in the section of the work under discussion entitled
A General Observation, Whiston confesses that he has been
unable to find a philosophical or mathematical solution to
the phenomenon of magnetism.   Just as Newton supposed that
gravity was derived immediately from the power of God, so
Whiston attributes magnetism to the same First Cause.[77]

The fact that his efforts to solve the longitude
problem by the dipping needle method were a failure does
not detract from his work as an intelligent and persistent
attempt to develop a method which had been suggested as
having great possibilities.

When George Graham published his papers on magnetic
dip in the Philosophical Transactions, Whiston was quick to
recognise that Graham's work invalidated his own work on the
longitude.

Commentaries on this 1721 publication of Whiston rarely
mention the ideas in Section XXIII with its 23 separate
corollaries.   Whiston calculates the rotation of the
magnetic pole to be about one degree in 5 years or an entire
revolution in 1920 years.[78]

Whiston is aware of Halley's 'four-pole hypothesis'
of terrestrial magnetism,

> "... so that if this exterior shell of earth be a
> magnet, having its poles at a distance from the
> poles of diurnal rotation; and if the internal
> nucleus be likewise a magnet, having its poles in
> two other places distant also from the axis, and
> these latter by a gradual and slow motion, change
> their places in respect to the external, we may
> then give a reasonable account of the four
> magnetic poles, as likewise changes of the needles
> variations, which till now hath been unattempted." (79)

Halley refers to the external globe as a shell which is separated from the internal globe by a fluid medium. Realising that this theory of an internal nucleus of the earth will rouse opposition, Halley draws the comparison with the globe of Saturn surrounded by its ring but both globe and ring having a common centre.

Whiston builds on these theories of Halley that

> "... this diurnal motion was imprest on the
> Earth, being given to the external parts, and
> from thence in time communicated to the
> Internal." (80)

He calculates that the ratio of the number of days in the upper earth's diurnal revolution, to the number of days in the revolution of the internal magnet is as 1 : 700,000. As already postulated in his New Theory of the Earth, Whiston assumes that the present diurnal rotation of the earth accelerated to its present rate of rotation from rest. In the same way the rotation of the internal shell of the earth was transmitted by the intervening fluid from the outer shell. Whiston uses these theories of the relative rotation of the inner and outer earth to substantiate the views expressed in his New Theory of the Earth about the age of the earth as calculated by biblical chronology.

> "... For if the world has been a very great number
> of thousands of years old in its present state,
> and without any great change by a deluge, we shall
> find that the old year must have been 1000 days
> long then 500, then 400 ... and at last only $365\frac{1}{4}$;
> and the months in proportion contrary to all the
> ancient accounts, records, or traditions sacred
> and prophane." (81)

and later he expresses his satisfaction as follows:

"... it is plain that these phenomena of nature,
as compared with ancient history, greatly
confirm what I have long since supposed and
almost demonstrated from other evidence in my
New Theory viz, that there was above 4000 years
ago a general deluge of water on the earth, and
that by such a cause as did lengthen the year;
and did also answer to all the description of
that flood in Moses History and to the other
phenomenon of nature and of antiquity." (82)

Whiston speculates about the nature of the fluid
separating the external shell of the earth from its internal
core. "Perhaps therefore that fluid may prove such as
is fit for animals to live in." In this, he is however
echoing Halley's own reflections on the nature of this fluid.

It is worth noting that Whiston aims at a unity of
thought in his studies on terrestrial magnetism and biblical
chronology, his conclusions being based on common scientific
principles applied to both areas of natural philosophy.

3.7 DETERMINATION OF LONGITUDE BY THE DURATION OF SOLAR ECLIPSES

In 1724 there was heightened interest in the phenomenon of solar eclipses due to the anticipation of the total
eclipse of May 11th in that year. Whiston published the
tract, The Calculation of Solar Eclipses without Parallaxes
(1724) as discussed in Chapter 4. He makes a Proposal
(pages 74-82) For the Discovery of the Longitude of the
several places of the Earth, by total eclipses of the sun.

"It is humbly proposed that observations be made
in all places where solar eclipses are seen, of
the exact duration of the same; by either viewing
the beginning and ending thereof through a telescope, with a glass smoked in the flame of a
candle ... or else by casting the sun's image
through such a telescope upon white paper, and

viewing the first and last impression of the
moon's shadow upon it."   (83)

Whiston gives a list of nine solar eclipses and
seventeen lunar eclipses visible from London between the
years 1724 and 1740, taken from the calculations of
Mr. Leadbetter.   Earlier in the same work, in Problems XIV,
XV and XXV, Whiston gives sample calculations for the solar
eclipse of May 11th 1724 for three places, London, Paris
and Dublin, and shows how it is possible, knowing the
latitude of the place and the duration of the eclipse, to
calculate the longitude of the place.

There is no doubt that this proposal offers a method
of determining longitude given the conditions specified,
but it did not really contribute to the solution of the
problem.   What was required was a method which could be
used anywhere and at any time and was independent of
weather conditions.

3.8   WHISTON.   LONGITUDE DETERMINATION FROM JUPITER'S
       SATELLITES

In 1715, Whiston refers to the possibility of finding
longitude by the eclipses of Jupiter's satellites.

"among the celestial methods, that is the Principal,
and hitherto hath been accounted such which is
grounded upon the eclipses of the satellites of
Jupiter."

. . . .

"... but as for the other (method of Jupiter's
Satellites), it is a most excellent method indeed,
and most sufficiently adapted to the sailor's use,
as well as at Land, IF BY ANY MEANS TELESCOPES
COULD BE USED AT SEA.   Accordingly we must here
explain it ... it is manifest that the moments of

the occultations and egress of these satellites
may be defined and determined for any meridian
whatsoever, and that those eclipses are real
phenomena, affixed to certain moments of absolute
time which may be observed in divers places of
the earth; nay they not unfrequently happen every
day, and so may serve to your purpose daily, if you
be provided with a computation of the eclipses of
them all, as it fit you should ...   There being
given therefore by computation the moment of
immersion and emersion, accommodated to the places
of the tables; and there being given at the same
time, by observation, in another place, whether by
sea, or in some distant country, the same moment;
there will also be given the difference of time:
which being turned into degrees and minutes of the
equator, shows the difference of meridians, or the
longitude of the place.   This is that celebrated
way of searching out the longitude, which astronomers,
the French especially, have so often practised, for
reforming the geographical situation of places,
and perfecting charts, whether of sea or land." (84)

Whiston gave up his work with dipping needles in 1723
and as early as 1730 had given considerable thought to the
design of a telescope for use at sea in the observation of
Jupiter's satellites.

"October 22nd 1730, I laid before this Society a
newly invented sort of refracting telescope
having seven object glasses to one eye-glass ...
I then observed that such a telescope would shew
the eclipses of Jupiter's satellites at sea, and
consequently the longitude there, notwithstanding
the rolling of the ship; that motion only removing
the planet from one object glass to another, but
still exposing it to the eye in one situation as
well as in the other.

My specimen there produced was nine foot
long, the eye glass was about 2" broad ...   The
instrument became capable of improving the discovery and sight of Jupiter's planets 28 times." (85)

Whiston asks for a ship to be appointed to go from Great
Britain to New York to test the suitability of this method
for determining the longitude within the limits of the Act
of Parliament.   He describes in detail the "contrivance
here made use of for avoiding the motion of the ship",

which was in part communicated to him by Mr. Bucknall on
Sept. 14th 1730.

> "Take a strong circular table of about 6 foot
> diameter, near its centre, a large annular circle
> of lead of 100 pounds weight or more be firmly
> fixed.   In the very centre of this table but
> about 6 inches on the upper surface of the lead
> the half of a small (square) concave socket in
> steel and under it let a very strong rod of
> steel about two feet long be affixed to any part
> of the ship above deck, but as near to the main
> mast as convenient will allow in perpendicular
> situation whose top is to be formed into a
> small convex hemisphere and to be inserted into
> the former concave hemisphere as a ball to its
> socket.   These are to be made as small and glib
> as possible.   This table when once adjusted to
> a proper vertical and horizontal situation will
> always keep it very nearly.   In the centre of
> the table above,erect a perpendicular support to
> be raised and depressed as occasion serves with
> a notch for a telescope and let the end near the
> eye be made much heavier than that towards the
> object, then so the centre of gravity which must
> ever be in the notch and its situation preserved
> by a screw) may be nearest the eye during the
> observation.   When anyone goes upon this table
> to make an observation, let him place a piece of
> lead equal to his own weight diametrically
> opposite to him and at the same distance from the
> centre as a counter poise.   When things are thus
> prepared it appears both by former and present
> trials that in ordinary weather at least the
> vibrations of the table and telescope will not
> be so much as the angle taken in by telescope
> which may be about five degrees and by conse-
> quence it appears that every eclipse of any
> satellite will be here in a manner, as visible
> at sea as at land and will discover the longitude
> with equal accuracy.

> NB.   The compass also had better by hung on such
>   a single centre than on Gimballs.   (86)

On 7th November 1734, Whiston again addressed the
Royal Society with 'a new discovery he hath since lit upon'

> "He proposes to have a reflecting telescope of
> Mr. Cassegrain's form, the tube of which is to
> be about four feet in length, with an aperture to
> the principal reflector of four inches and a half,
> but its focal length to be eight feet ...   he

proposes, that there should be nineteen eye-
glasses fitted, that is eighteen placed in two
circles about one in the centre, and the focal
length of each eye-glass to be about half an
inch ... he proposes also that it shall be
fitted with cross-hairs and a spirit level,
directing the sight to three or four degrees
above the horizon." (87)

Two interesting comments on Whiston's proposal follow

in the Journal Book of the Royal Society. Richard

Graham said

"he himself did some time ago propose to several
members of the Society who were now present,
something like one of the contrivances mentioned
in this description."

James Hodgson (nephew of Flamsteed) who was present

remarked,

"that the laws of the motions of the satellites
of Jupiter are not yet brought to that degree of
exactness as is sufficient to ascertain the
place of the innermost satellite, which is the
most regular of the four, to within four
minutes of time."

It was not until 1749 that Hodgson published his work,

Theory of Jupiter's Satellites with the construction and

Use of Tables for Computing their Eclipses.

On 12th August 1737, Whiston delivered to the

Admiralty Office his memorial relating to the determination

of longitude by the observation of Jupiter's satellites and

the reflecting telescope made by Mr. Chaplain. (88)

In 1738, Whiston addressed a further printed work to

the Commissioners appointed by Act of Parliament for the

discovery of Longitude at Sea. This work was entitled

The Longitude Discovered by the Eclipses, Occultations and

Conjunctions of Jupiter's Planets (1738). The same work

included <u>Descriptions of those refracting and reflecting</u>
<u>telescopes; and of those sectors and that quadrant, which</u>
<u>are the instruments necessary for this discovery, both at</u>
<u>Land and Sea</u>.

Also included was <u>An Ephemeris for the latter half of</u>
<u>the year 1738 containing the Configurations of Jupiter's</u>
<u>planets at six o'clock every evening while Jupiter is to be</u>
<u>seen.   With those eclipses, occultations and Conjunctions</u>
<u>that are useful for the discovery of the Longitude both at</u>
<u>Land and Sea</u>.

In 1741, Whiston wrote an Historical Preface* to his
work on Longitude, <u>The Longitude Discovered by the Eclipses,</u>
<u>Occultations and Conjunctions of Jupiter's Planets</u> (1738).
In this preface he gives us a full account (62 pages) of
his work on Longitude.

There is at present no evidence available that
Whiston's work on the determination of longitude by Jupiter's
satellites was put into practice at sea.   Enquiries to
instrument repositories and museums have not brought to
light the refracting and reflecting telescopes made for
this purpose.

There was a shift of interest to the development of
chronometers and in 1736, John Harrison had already tried
out his Number 1 timepiece on the Humber.   It was not until
1765 that the Board of Longitude finally recognised that
Harrison's Number Four timepiece had fulfilled the requirements for the prize of £20,000.

* The Historical Preface was not published separately but
is found bound to some copies of the 1738 publication.

3.9 PREPARATION OF A COASTAL SURVEY

This final and more successful effort which Whiston
made to solve the longitude problem was the initiation of
a new coastal survey.   At nearly 73 years of age, Whiston
still took an active interest in this project and with
others drafted the following proposal addressed to the
Commissioners for The Discovery of the Longitude at Sea.

> "The petition of Edmond Halley, William
> Whiston, John Machin, James Bradley and
> Thomas Haselden, Mathematicians"
> (not dated but most probably dating from 1739)

Edmond Halley (1656-1742) had been Astronomer Royal
since 1720 and was now 83 years old.   John Machin was the
originator of a theory of the moon's motions which had been
published as an addendum to the first edition of the
Principia to appear in English translation (1728).   James
Bradley (1693-1762) was to succeed Halley as Astronomer Royal
in 1742.   Over a period of twenty years (1727-1747) he had
made a series of observations at his private astronomical
observatory at Wanstead.   Thomas Haselden had been for
twenty years a schoolmaster on H.M.'s ships and later kept
a school at Wapping.

The proposal stated:

> "The principal thing which is absolutely necess-
> ary ... is the exact determination both of the
> latitudes and longitudes of the several ports
> whence and whither the ships are to sail, and
> before which determination the discovery of
> the longitude itself at sea, by what method
> soever it be attempted would be of very little
> benefit, nay sometimes would be really danger-
> ous to Navigation."

There were heavy demands for financial support.

"We therefore humbly represent that if your
Lordships please to allot us a ship for our
preparatory trials and to permit us the use
of the entire sum of £2000, we have great
reason to believe that we shall be able, in
two or three years time, to make and to procure
so great a number of good observations as will
be sufficient for such exact determinations of
the latitude and longitude of all our ports as
is necessary, by way of preparation for the
discovery of the Longitude itself at sea."   (89)

Another draft proposal,

"... For the Discovery of the Latitude and
Longitude both at Sea and Land"

probably dating also from 1739 is more explicit in its

suggestion of how the £2000 is to be spent.

"We humbly propose to your Lordships

I.   That out of the £2000 allowed by the Act of
      Parliament, Dr. Halley at Greenwich and Mr.
      Bradley at Wanstead, with telescopes of equal
      length and goodness be employed to observe,
      for two or three years together, all the
      visible Appulses of the moon to considerable
      fixed stars and all the visible eclipses of
      Jupiter's planets.

II.   That during the same years, the best observers be
      employed on board the ship to be allotted for this
      purpose, to sail round the coasts of Great Britain
      and Ireland; and to the several British Islands
      in the way to and along the coasts of our
      American plantations; and in the several ports
      with telescopes of equal length and goodness with
      those forementioned, to make the like observa-
      tions both of the appulses of the moon to the stars
      and of the eclipses of Jupiter's planets, as also
      of the intervals between the rising or setting of
      two fixed stars and between the rising or setting
      of fixed stars and the moon: and that the East
      India Company be desired to procure the like
      observations in the ports anyway belonging to their
      voyage and settlement in the East Indies.

III   That in order to the earliest advantages that may
      be made of these observations each observer be
      obliged to transmit every quarter of a year a
      catalogue of his observations, made the last quarter
      to the Lords of the Admiralty, that by a comparison
      of them with one another, the latitude and longitude
      of such ports where observations have been made,
      may be immediately determined.

IV   That in order to the earliest application of these
methods to the discovery of the longitude at sea,
Dr. Halley be allowed, as soon as possible, to
deposit with the Lords of the Admiralty his own,
and Mr. Hodgson to deposit Mr. Flamsteed's
Astronomical Tables; and that Mr. Machin be
allowed to do the same with his Calculation
of the moon's place, from his improvements of
Sir Isaac Newton's Theory of the Moon. Each of
them receiving out of the £2000 the sum of [not
specified] immediately, upon their depositing the
same Tables and Calculation.

V   That in order to the still greater advantage of
the publick, the persons employed to make the
observations abroad be enabled out of the £2000
to purchase and take with them such other instruments as may be useful for the Trial of other
methods of discovering the Latitude or Longitude;
or for making other experiments relating to
Natural Philosophy, Natural History, Astronomy
but especially to Navigation: such as long Plumb
lines, Loadstones, Horizontal and Dipping Needles,
Pendulum Clocks or Watches, Cross staffs, Mr.
Davies's Quadrant, Captain Elton's Quadrant [see
photograph], with and without telescopic sights
and that they be directed to make the best observations with them and send their quarterly observations as aforesaid." (90)

The above extract from Whiston's draft proposal
represents a comprehensive appraisal of the state of the
Longitude problem in England in 1739. Although there is
no specific mention of Harrison's chronometers, Whiston
does refer to the use of good pendulum clocks and watches.

The two final proposals of Whiston exist in manuscripts dated November 1739 and June 20th 1740. Both
petitions did in fact reach the Board of Longitude. In
the first of these documents, Whiston again stresses the
urgency of charting the coasts.

"That at the same time, new and accurate charts
may be made at our sea coasts for the advantages
of both Geography and Navigation",

and urges the compilation of a new world map of Variation.

"That Dr. Halley made a most useful map of the
Variation of the magnetick needle fitted to the
year 1700. Which variation is yet so greatly
altered since that time, that the same map may
now greatly mislead such mariners, as do not
carefully allow for that alteration. Which
variation may be anew exactly observed on this
ocean at all our ports.... Other observations
made since 1700 by Captain Middleton and Captain
Morton and many others may be compared in order
to the forming of a new map of that variation
for the year 1740." (91)

In the same document, Whiston refers to John Harrison's*
clock. It was George Graham who had forwarded £200 to
Harrison to enable him to make his first timepiece.

The memorial of June 20th 1740 is addressed to "The
Right Honourable, the Commissioners for exercising the Office
of the Lord High Admirable of Great Britain." Whiston
points out that the petition of 24th November had been
examined by the Commissioners appointed by Act of Parliament
for the Discovery of the Longitude at Sea, and that the
request concerning a new map of the variation of the magnetic needle had been accepted and was already being undertaken. The above mentioned Commissioners had evidently
directed Whiston to apply to the Admiralty for financial
support in his work surveying the coasts.

---

* Harrison had spent the years 1729-1735 in building his
first marine time-keeper. He successfully tested it on
board a barge in the Humber and brought it to London in
the spring of 1736. On the Royal Society's recommendation, the Admiralty allowed **Harrison** to embark with
his timepiece on the H.M.S. <u>Centurion</u> for a voyage
to Lisbon. As a result of this successful trial, the
Board of Longitude began to advance small sums to
Harrison from time to time. As it happened, Harrison's
first time piece fulfilled the requirements of the Board
of Longitude and was worthy of the full £20,000 reward.
(92).

"He now applies himself to this Honourable Board
and requests that they would please to take the
scheme he now lays before them into their serious
consideration and that as soon as possible; that
the advantageous position of Jupiters planets, the
eight following months beginning with August, may
not be neglected. Your petitioner also begs
of you to consider further, that since the present
designs of settling the coasts of Great Britain
and Ireland may be perfected in those eight
month's time, as appears by the computations
annexed, for about the sum of £800." (93)

Whiston did finally obtain a grant of £500 towards instru-

ments for this purpose as is recorded in the Minutes of the

Board of Longitude.

"At a meeting of the Commissioners for the Discovery
of the Longitude at sea on 16th January 1741
<u>Present</u>
Sir Charles Wager - First Lord Comm. of the Admiralty
Lord Monson
Sir Thos. Hanmer
Sir John Norris - Adm. of the Fleet
. Folkes Esq., P.R.S.
Dr. Smith. Prf. Astr. Cambridge
Dr. Bradley. -do-   Oxon.

Mr. William Whiston representing that he hath procured a new set of astronomical instruments for
finding out the Longitude and Latitude of the Coast
of this Kingdom with the variation of the needle,
and that having been at a considerable expense
already in the preparing of the said Instruments
and in purchasing divers things necessary for the
better enabling him to make observations with them
as also in making such observations on the Coast
at several times by himself of the persons by him
employed appears by an account which he delivered
under his hand, And that he shall not be able to
proceed in completing his said observations along
the coasts and as far as Cape Clear in the Kingdom
of Ireland unless he is supplied with the sum of
£500: And the Board being of Opinion that the said
Instruments may tend very much to the Advantage of
Navigation and that careful experiments thereof
should be from time to time made.

Resolved to give the said Mr. Whiston the sum
of £500 to enable him to make such experiments
upon Condition that he shall not without particular
orders for so doing, put the Public to any further
expense on any account whatever in relation to his
said Instruments in the making of any Trial or
Trials of the same." (94)

On February 23rd 1742 Whiston wrote to his daughter
Mrs. Samuel Barker at Lyndon, Rutland and told her his
good news.

> "You will soon see in print an account of my
> obtaining the £500 I desired of the Commissioners
> for myself ...." (95)

Whiston employed Mr. John Renshaw as his agent in
making the survey of the coasts,

> "Who surveyed it trigonometrically from the
> North Foreland in Kent to Land's End in
> Cornwall and the Scilly Island." (96)

During the year 1740-41, Whiston's work on the coastal
survey continued and was financed by the subscriptions of
Sir Charles Wager, first Commissioner of the Admiralty, the
Duke of Cumberland, Lord Wilmington, Lord Baltimore and
Mr. Townsend who together raised £175. (97)

In the same place Whiston gives a copy of the directions
given to Mr. Renshaw and Mr. Birkbeck who were employed in
making the coastal survey. Two copies of this large chart
815 x 1670 mm. are held in the Map Room of the British
Museum. It is entitled, <u>An Exact Trigonometrical Survey</u>
<u>of the British Channell from the North Foreland to the</u>
<u>Scilly Islands and Cape Clear on the Southwest part of</u>
<u>Ireland</u> (1743). (98)

The map was on the scale 1 : 550,000 and has five
insets drawn to a larger scale of the Islands of Scilly,
Falmouth, Plymouth, Isle of Wight and the River Thames.
Notes on the map are reproduced below.

1. Partly by private benefactions, but principally
    by the Publick Money assigned for this purpose by

the Commissioners of the Longitude. Being a correction and improvement of Dr. Edmund Halley's Chart of the Channel between England and France, first published in the year 1699.

2. The variation in the needle has been westerly in all parts of the Channel, since about the year 1657 and is now 16°41, therefore if you allow a degree variation for every 5½ years you will always have the variation nearly in the Channel.

3. This chart is sold by Mr. Whiston and Mr. Renshaw in Davis Street near Grosvenor Square. As also by Mrs. Senex over against St. Dunstans Church, Mr. John Whiston both in Fleet Street and Mr. Mount on Tower Hill. Price Six Shillings.

4. Advertisement

Land survey'd and mapt, and all parts of Mathematicks taught by the Author at his house in Davis Street Grosvenor Square, and to be heard off at Mrs. Senex's at the Globe or at Mr. Millward's at the Dial and Three Crowns both in Fleet Street, London. Where this Map is sold as also by Mr. Hikey at Bristol and Mr. Greaves at Portsmouth.

During the year 1740-41, Whiston revived his earlier ideas about the use of shells or balls of fire to ascertain longitude. Mr. Lynn of Southwick was one of Whiston's land observers who made careful records of the motions of Jupiter's satellites. In his letter of November 10th 1740, he informed Whiston of an idea he had about the possibility of determining Longitude by comparing observations made on shooting (falling) stars or meteors. Whiston tells us,

"It came into my mind that we might have a much better succedaneum than that proposed by Mr. Lynn for distances not too great by making such explosions of shells or balls of fire." (99)

Whiston quotes a series of results with mortars fired at Woolwich on 26th July 1738 by Brigadier Armstrong, and notes that if a mortar be fired 5° from the perpendicular, the horizontal range of the shell will be almost a third part of its altitude.   The Preface to his 1738 publication

ends with a map of the distances of signals from mortars.

The map shows,

"The limits of the distinct view of balls of fire
to be thrown up a quarter, and a half, and an
entire Mile high from Shooter's Hill near
Greenwich out of 6, 8 and 10 inch mortars;   as
they may be seen from any high hill, or the top
of the mast of any great ship.   These limits
are the first 40, the second 60, and the third
80 Geographical miles from the mortars."

Experiments made at Woolwich out of the brass 13"
sea Mortar, with the wooden bed, in the Bomb room
26th July 1738 by Brigadier Armstrong.   (100)

No. of Shells	Weight of Shells		Elevation		Wt. of Powder		Distance	Time of Flight
	Empty	Filled with sand	D	M	lb.	Ounce	Yards	Seconds
	lb.	lb.						
1	240	245	45	0	30		4100	....
2	230	240	45	0	26		3600	25
3	226	235	45	10	30		3627	$25\frac{1}{2}$
4	222	231	45	15	30		3580	$24\frac{1}{2}$

This account shows that for thirty years, 1713-1743,

Whiston maintained a lively interest in the longitude problem.

His achievements include the initiation of the 1714 Act of

Parliament;   his chart of the English Channel published in

1743 was one of the best available for many years.   No better

chart was produced until the major work of the Ordnance

Survey which began in 1747, and over the next sixty years

established an accurate trigonometrical framework for the

whole country.

CHAPTER 3 - REFERENCES

1. May, Commander W.E. Alexander Neckham and the Pivoted Compass Needle. Journal of the Institute of Navigation, Vol. VIII, No.3.

2. Ibid.

3. Mitchell, A. Chrichton. Chapters in the History of Terrestrial Magnetism. Terrestrial Magnetism and Atmospheric Electricity. Vol. 42, No.3, p.244.

4. Thompson, Silvanus P. Petrus Peregrinus de Maricourt and his Epistola de Magnete. Proceedings of the British Academy, Vol. II, (1906).

5. Formaleoni. Saggi sulla nautica antica de Veneziani Venice (1783).

6. Taylor, E.G.R. The Mathematical Practitioners of Tudor and Stuart England, C.U.P. (1967) p.56.

7. Waters, D.W. Galileo and Longitude, Physis, Anno VI (1964) Fasc 3.

8. Waters, D.W. Time, Ships and Civilisation. Antiquarian Horological Journal, June 1963.

9. Op. cit. (reference 7).

10. Ibid.

11. Gellibrand, H. A Discourse Mathematicall on the Variation of the Magnetic Needle together with its admirable diminution lately discovered. (1635).

12. Op. cit. (reference 6), p.232.

13. The Old Royal Observatory. National Maritime Museum publication (1960), p.4.

14. Marguet, F. Histoire Générale de la Navigation du XVe au XXe siècles

15. National Maritime Museum. Photograph A.6215.

16. Op. cit. (reference 6), p.124.

17. Whiston, W. Historical Preface to The Longitude and Latitude found by the Inclinatory or Dipping Needle, second edition (1721), ix.

18. Williams, A.C.   Ph.D. Thesis (1940), University of London.

19. Gould, R.T. The Marine Chronometer (1923), p.2.

20. Guardian, No. 107, July 14th 1713.

21. Parliamentary Debates, 1713-1714, p.144-145.

22. Commons Journal, Vol. 17, p.677-678, 11th June 1714.

23. Royal Society MS. MM 7.7

24. Gould, R.T. John Harrison and His Timekeepers. National Maritime Museum (1959), p.3-4.

25. Ibid., p.15.

26. Biographia Britannica (1766), p.4210.

27. Op. cit. (reference 19), p.5.

28. Op. cit. (reference 26), p.4210.

29. Trinity College Cambridge.   Dawson Turner Collection. MS. 164.   Cotes to Whiston December 2nd 1714.

30. Ibid.   MS. 166.   Whiston to Cotes, April 1715.

31. Field MS. F102.   Whiston to Samuel Barker, January 1715.

32. R.G.O. F 36/128.   Derham to Flamsteed.   November 3rd 1707.

33. R.G.O. F 36/143.   Derham to Flamsteed.   July 16th 1707.

34. R.G.O. F.34/119.   Sharp to Flamsteed.   1714.

35. R.G.O. F 34/114.   Sharp to Flamsteed.   April 3rd 1714.

36. R.G.O. F 37/150.   Broadsheet on Floreo; May 1715.

37. Ibid.   MS notes by Crosthwaite.

38. R.G.O. F 37/165.   Young to Flamsteed, 1714.

39. Whiston, W. A New Method for discovering the Longitude both at Sea and Land, Second edition (1715).   Introduction, p.18.

40. Ibid. p.25.

41. Philosophical Transactions No. 313, January 1708.

42.   Op. cit. (reference 39), p.103-104.

43.   Hodgkiss, A.G.   Discovering Antique Maps.   Shire
            Publications (1971).

44.   R.G.O. P 36/131.   Derham to Flamsteed, 1703. 29ᵗʰ Sept.

45.   Waters, D.W.   The Art of Navigation in England in
            Elizabethan and Early Stuart Times (1958)
            p.247-248.

46.   Op. cit. (reference 6), p.176.

47.   Gilbert, W.   De Magnete (1600).   Translation.
            Dover Publications Inc. (1958).

48.   Op. cit. (reference 45).   Plate LXIII.

49.   Purchas, Samuel.   Purchas his Pilgrimage (1613),
            Vol. III.

50.   B.M. Add MS. 4393 F.38.

51.   Op. cit. (reference 17), Preface (vi).

52.   Philosophical Transactions.   No.292

53.   Op. cit. (reference 17), Preface (xxiv)

54.   Ibid.   Preface (xxvii)

55.   Hellman, G.   Neudrucke von Schriften und Karten
            Vol. 4 (1895).

56.   Op. cit. (reference 17), p.53.

57.   Taylor, E.G.R.   The Mathematical Practitioners of
            Hanoverian England 1714-1840 (1966),
            p.135-136.

58.   Op. cit. (reference 55), p.13.

59.   R.G.O. 69. 192V (Herstmonceux).   Whiston to Flamsteed,
            November 25th 1718.

60.   Royal Society (Journal Book Copy) J.B.C. 12. p.32-35.

61.   Ibid.

62.   Ibid.

63.   Royal Society (Letter Book Copy) L.B.C. p.204,205.

64.   Royal Society, J.B.C. 12, p.49,50.

65.   Royal Society, J.B.C. 12, p.219.

66. Op. cit. (reference 26), p.4210.

67. Whiston, W.   Memoirs of the Life and Writings of Mr. William Whiston (1749), p.296-297.

68. Whiston, W.   The Calculation of Solar Eclipses without Parallaxes (1724), p.87-88.

69. Ibid. p.89.

70. Whiston, W.   The Longitude Discovered by the Eclipses, Occultations and Conjunctions of Jupiter's Satellites (1738), Preface (xi).

71. Op. cit. (reference 68), p.87.

72. Philosophical Transactions (1720-1723) abridged by Reid and Gray.   Vol. VI, Part II, Ch. IV, p.187-189.

73. Op. cit. (reference 17), p. 10.

74. Philosophical Transactions No. 344.

75. Op. cit. (reference 17), p. 19.

76. Ibid., p.11.

77. Ibid., p.88.

78. Ibid., p.57-58.

79. Philosophical Transactions.   Lowthorp Abridgement. Vol. 2, p.616-617.

80. Ibid.

81. Op. cit. (reference 17), p.63.

82. Ibid., p.64.

83. Op. cit. (reference 68), p.74.

84. Whiston, W.   Astronomical Lectures read in the Public Schools at Cambridge (1715), p.244-246.

85. Royal Society (Record Book Copy), R.B.C. 19.1. p.156.

86. Field MS.   F 117, Sept. 27th, 1730.

87. Royal Society, J.B.C. 15, p.19-21.

88. B.M. Printed Leaflet 533. 1. 16.11.

89. Field MS.   F 100B. 1739(?)

90.   Ibid.

91.   Field MS. F 133. November 24th, 1739.

92.   Quill, H.   _John Harrison_ (1966).

93.   Op. cit. (reference 91).

94.   R.G.O.   _Minutes of the Board of Longitude_, Vol. V, p.3.

95.   Field MS. F 133A.   February 23rd, 1742.

96.   Op. cit. (reference 26), p.4212.

97.   Op. cit. (reference 70), Preface (xxvi).

98.   B.M. Map Room.   1068 (11) also K. Mar. III. 22.

99.   Op. cit. (reference 70), Preface (xxxv).

100. Whiston, W. _Praelectiones Physico-Mathematicae_ (1710) MS. notes in B.M. copy.   Shelf Number 874.1.4.

APPENDIX   I

B.M. Add MS 4393 F. 37.

(MS in Dr. Pell's hand)

The present state of our knowledge of the magnet seems to be this.

By Mr. Bonds Hypothesis and Tables of equal motions of the magnetic pole we may find the place of the magnetic pole in the triangle LMN where L signifies London, M, the magnetic Pole and N the astronomical North Pole of the same earths equator.   O some other place given by its longitude and latitude from London.   To calculate what respect a needle at O has to the point M; that is what variation it has and what variation it has and what inclination under its own horizon. And contra having the variation and inclination at O the position of O in respect of N & S of the Meridian NL.

Bond saith by his table of inclinations and on inclination after made (?) in any place of the earth's face and the place's latitude, he will find the longitude of the place. But he forgets that he must also know the time (or least the year) say not the hour of the observation.   For without doubt in another year, in the latitude of London, some other point far from London may have the same inclination that now London has.

I do not expect that an observation at sea can give me an inclination of the needle to 1/60 of a grade, but if it can be done on land it will serve to place all islands and prominences, and landmarks on the coast in their true places in a chart.   Having the longitude and latitude of Lands End and St. Helena, I can calculate the length of an arch of a great circle between Lands End and St. Helena.   I can calculate all the angles which the arch makes with the intermediall meridians.   But cannot by any means sail according to those angles (and thereby keep the via brevissima inter duo puncta proposition).   But if Bond teach me how to calculate all the intermediate variations in every point of that arch, I can turn my astronomical angles into magnetical angles and so prescribe them to my steersmen.

But I have reason to believe that no steersman can follow my prescription and therefore, I must have my new Fashioned Compass where I can place my needle as much awry every day as I believe the compass varies, and so the finger above shall always be kept to the line or marked point in the edge of the box as if the steersman always steered precisely

north.   But if he desire to name winds or other mark, he
must have another old-fashioned compass with all its winds.

37 (reverse)

Signed May 25th 1673.

1.   The friendship between magnets and iron or steele was
     known to Europe 2,000 years ago.

2.   The verticity of a magneticall needle was not known
     in Europe 3,000 years ago.

3.   There is no certainty that any compasses (Bussolas)
     were made before Flavio of Malfi in Italy.

4.   At the first they thought that every (compass) showed
     the true north.

5.   Next they found it showed true north nowhere.
     The swerving they called the variation.

6.   They thought the variation was different in different
     places but always the same in the same place, so
     always pointing to a magnetic pole in Zona Artica.

7.   That at London it was (11¼) one point of the compass
     eastward and would always be so.

8.   Mr. Burroughs set down in print what he found it at
     Limehouse near London A.D. 1580.   Linton's news of
     the complement of Navigation.

9.   Robert Norman first found the dipping of the needle
     Anno 1576.

10. Hereupon he wrote his treatise called The Newe
     Attractive.   To show that there was a point respective
     but not attractive of the needle.

11. Burrowes out of the Inclination and Variation calcu-
     lates the places of the respective point in the body
     of the Earth.

12. Dr. Gilbert makes a book 'De Magnete' and in it tables
     to find the Latitude by the magnet.

13. Mr. Gunter makes a diall in the Kings Garden at
     Whitehall.   Many mathematicians assist to find the
     meridian.   When it was set, they apply their
     compasses to it and the variation not 8 grades.

14. Mr. Gunter goes to Limehouse and finds the variation
     as he set it down in his book 1623 as I remember.

15. John Marr applies his compass to the same dial and
     finds it diminished.

16.   Mr. Gellibrand goes to Limehouse. He writes a book of the Diminution of the variation. 1633. John Pell writes an exercitation upon it.

Cabeus?

17.   Mr. Henry Bond about the year 1635 begins to examine the needle.

18.   Mr. Pell goes to Mayfield to speak to Mr. Gellibrand concerning Longitude 1636.

19.   Mr. Pell (?) comes to London. Visits John Marr at Richmond 1698.

20.   Athanasius Kircher writes his Artem Magnetica. Mr. Pell corresponds with Marsennes, sends him his exercitation de diminute variationis compass.

21.   Henry Bond in Tapps Seaman's Kalendar (about 1636) foretells that in the year 1657, the London variation would be decreasing (?). It proves so.

22.   It is now become 4 grades westward.

23.   Dr. Merret gets a clinitary which belonged to ... and had been made by Robert Norman, Compass-maker at Ratcliff. He brings it to Mr. Bond to mend.

24.   For Mr. Bond had before this discovered the motion of the magnetic pole upon the face of the earth and easily was able to calculate the variation and the Inclination of the Needle for London or any other point upon the sea, or earth, whose latitude and longitude are given. In 1672 he prints proposals and gives Dr. Pell a copy.

25.   He proposes himself willing to train in Dr. Pell.

26.   Dr. Pell acquaints the governors of Trinity House with the proposals. He leaves one in the hands of Sir Joseph (Jordan ?).

APPENDIX II

List of tracts relating to the Longitude problem published
1714-1758.

(This list does not claim to be a complete record of all
the tracts published. Many claims submitted to the Board
of Longitude were not presented in the form of a tract.
Notes have been added to indicate the method involved where
this is not already clear from the title.)

ANON, 1714, An Essay ... towards a new Method to show the
Longitude at Sea: especially near the dangerous shores.
Printed for E. Place at Furnivals Inn Gate, Holborn.
(Lighthouse to flash time signals on clouds).

B., R. 1714, Longitude to be found out with a newly invented
instrument by Sea and Land (improved lunar tables).

BILLINGSLEY Case, 1714, The Longitude at Sea not be found
by firing guns or by the most curious spring clocks or
watches. But the only true secret ... now humbly proposed
by C.B. (probably a pendulum clock)

BROWNE, Robert, 1714, Methods, propositions and problems
for finding the latitude and longitude at sea by celestial
observations and also by watches.

HALL William, 1714, A New and True Method to find Longitude,
London, 1714 (method by watch and dial).

HAWKINS Isaac, 1714, Essay for the discovery of the
Longitude at Sea. (To measure the tidal rise and fall of
sea level by means of a portable barometer).

HOBBS William, 1714, A New Discovery of the Longitude
(Universal Clock).

JACKSON Benjamin Habbakuk, 1714, Some new thoughts ... concerning a threefold motion of the Earth ... and facilitating
the Discovery of the Longitude. (Tides, magnetic compass,
ocean currents and winds, motion of the moon).

PLANK Stephen, 1714, An Introduction to the only method for
discovering the longitude, London, 1714. (To carry an
accurate watch which should be kept warm by fire).

RICCI Sebastiano, 1714, A New method proposed by Segnior
Dorothes Alimari to discover Longitude. (Precise
ephemerides for a selected prime meridian and precise
observation of the sun's azimuth and altitude).

THACKER Jeremy, 1714, The Longitude Examined. (Use of a
chronometer).

WHISTON William and DITTON Humphrey, 1714, A new method for

discovering the Longitude both at sea and land. (Sound and light Signals).

FRENCH John, 1715, A Perfect Discovery of the Longitude at Sea No. 627. Math. Practitioners. (Fire under a Compass needle).

KINDON Henry, 1717, Certain new Hypotheses of Fundamental Principles ... by means whereof the distance and longitude at sea as well as at land are discovered and determined. (new solar tables).

WHISTON William, 1719, The Longitude and Latitude found by the Inclinatory or Dipping Needle.

PLANK Stephen, 1720, Introduction to a True Method for the Discovery of Longitude at Sea (lunar distances).

GORDON George, 1724, A Complete Discovery of a Method of observing the Longitude at Sea (Jupiter's satellites).

JACKSON Henry M.T., 1727, Longitude and Latitude in conjunction by Sea and Land, by Day and Night demonstrated. (Moon's appulses to the fixed stars).

WILLIAMS Zachariah (1672-1753), 1755, An Account of an Attempt to ascertain the Longitude at Sea by an exact Theory of the Variation of the Magnetic Needle with a Table.

WHISTON William, 1738, The Longitude discovered by the Eclipses, Occultations and Conjunctions of Jupiter's Planets.

MOUNTAIN William and DODSON James, 1758, An Account of the methods used to describe lines on Dr. Halley's Chart of Terraqueous Globe showing the Variation of the magnetic needle about the year 1756. Their Application and Use in Correcting the Longitude at sea.

184

C H A P T E R   F O U R

WHISTON'S   OTHER   SCIENTIFIC   ACTIVITIES

4.1   Mathematical Textbooks for the Cambridge Schools.

4.2   Astronomical Lectures.

4.3   Sir Isaac Newton's Mathematical Philosophy more easily demonstrated.

4.4   Experimental Lecture Courses for the General Public.

4.5   The Copernicus.

4.6   Various Astronomical Phenomena Reported.

4.7   Maps and Charts of Astronomical Interest.

4.8   Astronomical Principles of Religion.

References to Chapter Four.

<u>Appendix</u>

Scientific Publications of William Whiston.

C H A P T E R   F O U R

WHISTON'S OTHER SCIENTIFIC ACTIVITIES

Whiston's first scientific publication, <u>A New Theory</u> <u>of the Earth</u> (1696) put him at the centre of one of the popular literary and scientific controversies of his times. Chapter Five of this work includes a discussion of Whiston's religious works, which shows that his interest in biblical chronology was a link between two fields which we recognise as distinctly separate, namely scientific and religious studies. Whiston set out to apply his astronomical and mathematical skills in the religious sphere, to correlate secular events with the fulfilment of scripture prophecies, and to establish a firm basis for the chronology of events in antiquity. It is, therefore, impossible to try to categorise any of Whiston's activities as purely scientific. The distinction adopted in this chapter is to separate those printed works and items of correspondence which Whiston would not have been able to write were it not for his interest in natural philosophy fostered during his years at Clare College and which later matured during his tenure of the Lucasian professorship of mathematics.

Chapter Two of this work has already dealt in some detail with Whiston's interest in cosmology and earth history.*

---

* David Kubrin in <u>Providence and the Mechanical Philosophy</u> (Ph.D. Thesis, 1968, Cornell) examined Whiston's theory of the earth and other similar works of the period which deal with the creation and dissolution of the world.

Chapter Three has shown that Whiston's active participation in the Longitude problem began in 1713 and continued for at least thirty years until the completion of
the trigonometrical survey of the British Channel in 1743.
In the present chapter, it is shown that Whiston's engage-
ment in other scientific activities was unbroken from his
first appointment as deputy professor to Newton in 1701 to
the year preceding his death 1751.   This implies that concurrently with his prolific output of religious publications
there was a continuous flow of published material of a more
specialised scientific character.   It appears that Whiston
drew a substantial part of the financial support for himself and his family from publications of this sort.

4.1   MATHEMATICAL TEXTBOOKS FOR THE CAMBRIDGE SCHOOLS

In 1701, Newton appointed Whiston as his deputy in
the Lucasian professorship of Mathematics at Cambridge.
There is evidence that Whiston resigned his living at
Lowestoft and Kessingland in Suffolk and in February of
1701 was giving a series of Astronomy lectures in the
Cambridge schools.[1]   A year later, Jan. 8th 1702,
Whiston was chosen professor in succession to Newton though
he did not officially assume his professorial status until
May 21st 1702.[2]   During the year before Newton's resignation (1702), Whiston appears to have fulfilled the teaching
commitments of the professorship and Newton allowed him the
full salary of the post.[3]

In March 1703, Whiston published a Latin edition of

Tacquet's <u>Euclid</u>.   The full title of this work is <u>Andreae</u>
<u>Tacquet Elementa Euclidea Geometriae Planae ac Solidae et</u>
<u>Selecta ex Archimede Theoremata</u>.

Further Latin editions of Whiston's adaptation of
Tacquet's Euclid appeared in 1710, 1722 and 1795.   The same
work was translated into English and appeared in several
editions, 1714, 1719, 1727, 1728 (Dublin) and 1753.   Latin
editions of Tacquet's Euclid were printed from Whiston's
text in Venice in 1737 and 1781.   The short title of the
English work is usually listed as <u>The Elements of Euclid</u>.
Whiston tells us that his own early interest in Mathematics
was fostered by Tacquet's Euclid.

> "Now it was the accidental purchase of Tacquet's
> own <u>Euclid</u> at an auction that occasioned my first
> application to the Mathematics, wherein Tacquet
> was a very clear writer."   (4)

Euclid's <u>Elements</u> date from about 300 B.C.[5] and
consist of thirteen books.   The first book sets out basic
postulates, common notions and is followed by a series of
48 propositions.   Further books deal with geometrical
algebra, geometry of the circle, the theory of proposition,
geometrical arithmetic, irrationals, solid geometry,
exhaustion, regular solids in a sphere.

Many Latin versions of the <u>Elements</u> appeared after
1533.   In 1654 Andre Tacquet, a priest of the Society of
Jesus published <u>Elementa Geometriae Planae et Solidae</u>.   This
work contained only the eight geometrical books arranged for
use in schools.   There is evidence of a further Latin
edition of Whiston's <u>Tacquet's Euclid</u> printed in Amsterdam
in 1725.   A note of this edition appears in the

Catalogue of the Libraries of Sir Thomas Burnet published
by T. Osborne in 1754.[6]

Whiston gives an historical preface on the growth of
Mathematical ideas.   The author of Advice to a Young
Student, Daniel Waterland, praises this preface and
recommends the study of Euclid's Elements in the first year
of the undergraduate philosophical curriculum at Cambridge.

> "Euclid may follow [Well's Arithmetic], or be
> begun at the same time as the former if your
> tutor reads lectures in it; otherwise let it
> alone until he does.   I shall not trouble you
> with reasons why I prefer Euclid to any other
> elements of Geometry as the most proper to begin
> with: see Mr. Whiston's preface to Tacquet which
> I entirely agree with." [7]

Whiston's Euclid followed Tacquet in compiling the
eight books as follows: Books 1 to 6 and Books 11 and 12.
The same text includes, again following Tacquet, selected
theorems from Archimedes.   Definitions and axioms are
followed by forty-six propositions which deal with the solid
geometry of the sphere, cylinder and cone.

The importance of Whiston's Tacquet's Euclid is its
clarity and simplicity which made it suitable as an introduction to plane and solid geometry.   In all probability,
Whiston recommended his edition of Tacquet's Euclid to his
students at Cambridge.   Isaac Barrow had published a Latin
version of Euclid in 1655, Euclidis Elementorum Libri XV
breviter demonstrati.   This included Book XIV which is a
supplement to Book XIII and is by Hypsicles (about second
century B.C.) and Book XV on regular solids which is thought
to be by a pupil of Isidore of Miletus (about 532 A.D.).[8]
Barrow's work was highly compressed necessitating abbreviated

proofs and using large numbers of symbols.   Whiston
evidently judged that a new edition of Tacquet's Euclid
would better meet the needs of his pupils.

With Newton's permission, Whiston published in 1707
Arithmetica Universalis.   Whiston tells us that this work
was nine years' lectures on Algebra which Newton had
delivered as Lucasian professor.[9]   This edition contained many errors and was in time superseded by Machin's
edition.

In the publication of both these works, Tacquet's
Euclid and Newton's Arithmetica Universalis, Whiston was
catering for his undergraduates at Cambridge.   Newton's
Algebra was recommended for study by those who had already
qualified as Bachelor of Arts and were not intending to
proceed to Orders.[10]

4.2   LECTURES   AS   LUCASIAN   PROFESSOR   -   ASTRONOMICAL
       LECTURES

Whiston published two books giving the texts of
lectures he delivered during his professorship.   Both were
translated into English and went through several editions.
Praelectiones Astronomicae Cantabrigiae in Scholis Publicis
Habitae (1707) first appeared in English in 1715 as
Astronomical Lectures read in the Public Schools at
Cambridge.   A corrected second edition of the English
translation appeared in 1728.

Whiston's second book of lectures, Praelectiones
Physico-Mathematicae Cantabrigiae (1710) gave rise to a

second Latin edition in 1726.   It was translated into
English in 1716 as <u>Sir Isaac Newton's Mathematic Philosophy</u>
<u>more easily demonstrated with Dr. Halley's Account of</u>
<u>Comets illustrated</u> (see Plate opposite).

The importance of these two books of lectures is that
they constituted a fundamental course on natural philosophy
incorporating Newtonian principles suited for undergraduates.
In 1707 Roger Cotes was appointed to the newly created
Plumian professorship of astronomy at Cambridge.   Whiston
and Cotes worked in close co-operation during the years
1707-1710.

> "Mr. Cotes and I began our first course of
> philosophical experiments at Cambridge in May
> 1707.   In the performance of which, certain
> hydrostatic and pneumatic lectures were com-
> posed; they were twenty four in number.   The
> one half by Mr. Cotes, and the other half by
> myself."   (11)

Roger Cotes' <u>Hydrostatical and Pneumatical Lectures</u>
did not appear in print until 1737 when they were edited
by Robert Smith, Master of Trinity College, Cambridge.   A
second edition appeared in 1747.   Smith remarks in the
preface to this work,

> "The method is general of teaching Philosophy by
> Courses of Experiments, (that is, of drawing
> general truths and conclusions from a select
> number of simple experiments first represented
> to our senses and then explained to our under-
> standings) it is now so much practised and
> approved of by the most eminent professors all
> over Europe ...
> He (Cotes) has rendered the discourses more
> entertaining and useful than ordinary, by
> enlivening the Science with a mixture of
> learning and the history of the inventions he
> treats of."   (12)

Sir *Isaac Newton's*

MATHEMATICK
PHILOSOPHY

More eafily DEMONSTRATED:

WITH

Dr. *Halley's* ACCOUNT of
COMETS Illuftrated.

Being Forty LECTURES Read in the
Publick Schools at *Cambridge.*

By WILLIAM WHISTON, M. A.
Mr. LUCAS's, Profeffor of the Mathematicks
in that Univerfity.

For the Ufe of the Young Students there.

In this *Englifh* Edition the Whole is Corrected
and Improved by the AUTHOR.

LONDON:

Printed for J. SENEX at the *Globe* in *Salifbury-Court*;
and W. TAYLOR at the *Ship* in *Pater-Nofter-Row.* 1716.

It has been said that under the direction of Richard Bentley, then Master of Trinity, Cotes and Whiston started the school of physical sciences for which Trinity College has been famous ever since.[13]

Whiston's Astronomical Lectures (English) (1715) gave the text of 31 lectures delivered between February 1701 and December 1703. The lectures are followed by sets of astronomical tables:

> Mr. Flamsteed's (corrected)
> Dr. Halley
> Mons. Cassini
> Mr. Street

During his student years at Cambridge, Whiston was taught the principles of Cartesian philosophy. It was only after his ordination in 1693 that he began to study Newton's ideas.

> "After I had taken Holy Orders, I returned to the College (Clare), and went on with my own studies there, particularly the Mathematics and the Cartesian philosophy which was alone in vogue with us at that time. But it was not long before I, with immense pains, but no assistance, set myself with the utmost zeal to the study of Sir Isaac Newton's wonderful discoveries in his Philosophiae Naturalis Principia Mathematica, one or two of which lectures I had heard him read in the public schools though I understood them not at all at that time." (14)

After 1696, Newton took up residence in London in order to assume duties as Master of the Mint. The following four or five years were marked by a heightened level of activity among eminent teachers at Oxford and Cambridge to organise Newtonian ideas into a form suitable for the teaching of undergraduates.

One of the first and most successful attempts was the

publication of David Gregory's Astronomiae Physicae et
Geometricae elementa, Oxoniae, (1702).   Whiston's
Astronomical Lectures which appeared in Latin in 1707 were
another work in the same tradition.   His first lecture
was an oration to the University of Cambridge and is not
printed in the English text of 1715.   The first topic
treated in his lectures was the figure of the earth.   The
earth is shown to be an oblate spheroid and a fairly full
treatment is given to the problem of the measurement of a
degree.   Richard Norwood measured an arc of the meridian
in 1635 and gave a value of 139 miles to two degrees, a
measurement less than half a mile in error.   Picard and
Cassini working in 1671 near Paris used a base line of more
than seven miles and their measurement of a degree was only
a few yards in error.

The text of the lectures goes on to deal with the
distance of the fixed stars.   The question of an annual
parallax of the fixed stars was in debate at the time.
Whiston quotes Huygens in his Cosmotheoros as asserting the
impossibility of determining accurately the distances of the
fixed stars due to the smallness of the annual parallax.*
Both in the first English edition (1715) and its corrected
version in 1728, Whiston does seem to give credence to the
claims of Hooke and Flamsteed to have detected annual
parallax.

The appearance of new stars and stars of variable

---

* The actual parallax of a fixed star was not finally
confirmed until 1839-1840 by the work of Friedrich
Wilhelm Bessel who made measurements on 61 Cygnus. (15)

brightness is discussed in lecture V. Whiston shows that
he is familiar with the works not only of English astronomers but quotes the opinions of Huygens, Tycho, Ricciolus,
Mercator, Bullialdus, Hevelius, as well as the view of
J.D. Cassini.

Lectures VI and VII discuss methods for measuring the
diurnal parallax of the sun and hence calculating the earth-
sun distance and the distance of all the planets from the
sun with their diameters and periodical times. A table
giving these data appears at the end of Lecture VII.[16]

A page of lecture notes by Thomas Morell who must have
either attended Whiston's lectures or read his book of
lectures gives salient facts from these early lectures.[17]
This data agrees closely with that in lecture VII of
Whiston's Astronomical Lectures.

"The sun 81,000,000 miles distant from the earth.
The diameter of the sun 800,000 miles.
The parallax of the sun scarce exceeds 1/6 of a minute.

Mercury		32	
Venus		59	
The Earth	distance	81	
Mars	from the sun	123	000,000 miles
Jupiter		424	
Saturn		777	

Mercury		4240
Venus		7906
The Earth	diameter	7935
Mars		4144
Jupiter		81155
Saturn		67870

	D	H
Mercury	87	23
Venus	224	17
The Earth	365	6
Mars	686	23
Jupiter	4332	12
Saturn	10759	7

In lecture VII, Whiston enlarges on the method of

determining solar parallax when Mars is in opposition to the sun.   This method of calculating the dimensions of the solar system is attributed to Cassini.   Whiston, however, recognises the important contribution made by Townley's micrometer.

Lecture VIII deals with other problems of solar astronomy.   Gassendi and Halley are given as authorities for determination of the moment of the solstice.   The mathematical constructions for determining the moment of the solstice from the shadows of the sun's rays are taken from David Gregory's text book, Astronomiae Physicae et Geometricae elementa, Book III, Prop. XI.

Lunar theory is explained according to Horrox, Opera Posthuma, London (1673), which included his Luna Theoria Nova. Lunar libration was observed by Hevelius,[18] and its causes explained by Newton.[19]

Lectures X and XI are devoted to methods of calculating the position of the sun and the moon for any given moment. The method of calculation of the sun's position is taken from Mr. Flamsteed's tables which are included as part of the Appendix to this work.   Horrox's theory of lunar motion is recommended in calculating the position of the moon. Newton's contribution to lunar theory is acknowledged but it is recognised that lunar theory is as yet incomplete.

> "... Sir Isaac Newton, from his skill in these
> things, which is indeed wonderful, hath since
> the making of these tables discovered the true
> causes of the lunar inequalities; he hath also,
> by way of specimen showed how most of them may
> be calculated.   But an entire theory of this
> planet, a priori, even he hath not yet been
> able to accomplish; and for what he has

accomplished no Tables have been yet made.   For
which reason we content ourselves at present
with Horrox's hypothesis, being ready to
embrace what is more perfect as soon as we shall
be so happy to attain unto it."   (20)

Lectures XII to XV deal with the theory of solar and
lunar eclipses.   The method of calculating solar eclipses
by a geometrical method of determining the appulses of the
moon to the fixed stars is attributed to Wren and Halley
and was later adopted by Flamsteed.(21)

At the conclusion of the section on solar and lunar
eclipses Whiston draws attention to the application of this
knowledge to Chronology.

"Seeing we have dwelt so long upon eclipses,their
most excellent and singular use in matters of
history and chronology ought not to be passed
over in silence ...   So likewise by the calculation of eclipses as compared either with the
observations of old astronomers or the relation
of historians, we examine the beginnings of eras
and epochs, the courses of cycles, the years of
Emperors, and Kings and the credit of historical
matters."   (22)

Lectures XVI to XXX are devoted to planetary astronomy.
In Lecture XVIII, Whiston gives a clear and explicit account
of Kepler's contribution to planetary astronomy.

Huygens' Cosmotheoros is the authority quoted for the
data relating to the satellites of Saturn.   Huygens'
theory was revised and corrected by Halley who compiled
Tables which are part of the supplement to Whiston's
Astronomical Lectures.

Cassini's calculations of eclipses of Jupiter's
satellites also form part of this supplement and Whiston
gives sample calculations.   In Lecture XXI Whiston discusses

several methods for determining the geographical longitude of places and mentions the successful application by the French of the eclipses of Jupiter's satellites to mapping.[23]

Olaus Roemer studied the motions of Jupiter's moons and observed in 1675 that the intervals between successive eclipses of a moon were regularly less when Jupiter and the earth were approaching one another. He used these observations to make a rough estimate of the speed of light.

> "... it will be reasonable to determine that light
> is propagated from the Sun to the Earth in less
> than half a quarter of an hour: which distance,
> seeing it is determined by the best astronomers
> to contain somewhat above eighty million miles,
> it gives us occasion to admire and even to be
> astonished at this prodigiously swift propagation
> of light." (24)

Whiston recognises the possibility of the fixed stars being placed at enormous distances from the earth.

> "by reason of the immense distance of some of them
> [the fixed stars] the space of time in which
> their light is propagated to us will be estimated
> not by Hours or Days but Weeks, if not Months
> also." (25)

Whiston was aware of Halley's work concerning transits of Mercury and Venus across the disk of the sun and the possible application of these phenomena to determine the earth-sun distance. In 1639 Jeremiah Horrox had observed a transit of Venus, and Hevelius posthumously published this account.[26] Halley calculated the transits of Venus between the years 918 and 2004.[27]

There was great anticipation of the next transit of Venus in 1761.

"The next transit of Venus will be in the year
1761, May 26, a little before six a clock in
the morning and in this transit she will be
distant to the south from the centre of the sun
not above 4½ minutes."   (28)

In 1748, when an old man, Whiston wrote to Martin
Folkes, President of the Royal Society, begging him to take
precautions that the transit of Mercury on 25th April 1753
and the transit of Venus on June 5th 1761 (new style
calendar), should be duly observed.(29)

Throughout, Whiston's lectures, continual use is made
of Street's Caroline Tables (1661), which were a revision
of the work of Kepler and Tycho Brahe, published in Ulm in
1627 as the Rudolphine Tables.   Street's tables were part
of the supplement to Whiston's astronomical lectures.
Certain errors were recognised in Street's Astronomia
Carolina but Whiston suggests,

"Nor will Mr. Street's error therein hinder us
from the use of his Tables, since we may safely
join Mr. Flamsteed's Tables of equating Time
with the Caroline Tables of the Planets." (30)

In Lectures XXVI to XXVIII, Whiston discussed the
geometrical calculation of the places of the planets, and
shows that he is conversant with Keplerian planetary
astronomy.   He notices the unsolved problem of dividing,
by a geometrical method, the area of an ellipse or a circle
in a given proportion.   The approximations of Ward and
Bullialdus are outlined and Whiston gives his own improved
method in Lecture XVIII.

The last two lectures, XXX and XXXI, are devoted to
an exposition of Newton's Theory of the Moon.

"We should forthwith proceed to the famous Sir
Isaac Newton's Philosophy ... which [he] hath
very lately composed out of Mr. Flamsteed's
observations and Dr. Gregory hath communicated
to the public." (31)

Whiston is here referring to a short tract on the

moon's motion written by Newton about 1700. It was first

published in Latin in 1702 by David Gregory in his

Astronomiae Physicae et Geometricae Elementa. These

lectures represent Newton's theory of the Moon as it stood

between the first edition of the Principia in 1687 and its

second edition in 1713. In fact, the full impact of

Newton's distinctive contribution to lunar theory is not

represented in these two lectures. Newton deduced the

"inequalities" in the motion from general physical principles

using the law of universal gravitation. Up to that time

astronomers worked empirically and continually adjusted

their schemes to make predictions of lunar positions agree

with actual observations.[32]*

---

* Bernard Cohen's reprint edition of Whiston's Astronomical Lectures, second edition (1728) (Johnson Reprint
Corporation, 1972) renders two services to scholars
which are worthy of mention. A list of lecture titles
has been compiled and the work also includes an alphabetical list giving page references of the 52 persons
mentioned in Whiston's work. In assessing the importance of Whiston's Astronomical Lectures, I agree with
Cohen that they represent the astronomical background
to Book 3 of the Principia and are a fair indication of
the instruction given in astronomy at Cambridge in the
early decades of the eighteenth century.

4.3 SIR ISAAC NEWTON'S MATHEMATIC PHILOSOPHY MORE EASILY DEMONSTRATED

Whiston's second series of lectures as Lucasian Professor, Praelectiones Physico-Mathmaticae (1710) was particularly intended for the use of young students at Cambridge.   Extracts given here will be taken from the first English edition, Sir Isaac Newton's Mathematic Philosophy more easily demonstrated 1716.   It covers lectures given from February 1704 to December 1708.

Lectures 1-12	February 1704 - December 1704.
Lectures 13-14	January 1705 - December 1705.
Lectures 25-30	April 1706 - December 1706.
Lectures 31-33	January 1707 - November 1707.
Lectures 34-38	April 1708 - December 1708.

The final two lectures were concerned with Dr. Halley's Cometography.   Halley's essay A Synopsis of the Astronomy of Comets is given in its entirety.

The purpose of the work is announced by Whiston in the first lecture.

> "For we are purposed to trace the steps of the great man [Newton] and to set forth his principal and most noble philosophical inventions in a more easy method;   that so we may bring that divine philosophy within the reach and comprehension of those who are indifferently perhaps exercised in the mathematics."   (33)

This series of lectures is recommended as suitable for inclusion in the philosophical studies of fourth year undergraduates, together with

Baronius	Metaphysics
Newton	Optics
Gregory	Astronomiae Physicae et Geometricae Elementa (34)

The first commentary on the _Principia_ was David
Gregory's _Notae in Newton's Principia_, circulated in manuscript but never printed.   Whiston's lectures represent
the first extended commentary to have been published.   The
sources of Newtonian ideas available to Whiston for this
series of lectures were the first edition of the _Principia_
(1687), Newton's _Opticks_ (1704) and the manuscript of
Lectiones Opticae given by Newton as his inaugural lectures
on assuming the Lucasian Professorship, later deposited in
the University Library according to the terms of the
Professorship.[35]

Whiston's lectures include direct quotations of considerable length from the _Principia_, and thus represent the
first published version in English of a major portion of
the _Principia_.   Whiston's own commentary is interspersed
with direct quotations from Newton;   separation of commentary
from quotation would require a line by line consultation of
the _Principia_.   Whiston's particular philosophical bent
is revealed by his introduction of a Cartesian definition
of matter.

> "Body or Matter is an extended substance, solid,
> or impenetrable, of itself merely passive, and
> indifferent to Motion or rest;   but capable of
> any sort of Motion whatever and of all figures
> and forms.   I call it a substance extended,
> because it possesseth some part of extended
> space..."   (36)

The text begins with an elementary treatment of Conic
Sections followed by a study of the laws of motion.   The
lectures go on to discuss motion in a circle, elliptical
motion and projectile motion, and the theory is applied to
the motion of the planets and of the satellites of Jupiter

and Saturn.[37]

The motion of two bodies around a common centre of gravity is considered,

> "If two bodies drawing each other by any force whatever, and which are not moved from anything else, nor impeded, be moved in any sort whatever; their motions will be the same in effect as if they did not attract each other, but they were both attracted with the same force by some third body placed in the common centre of gravity." (38)

The variation in eccentricity of the lunar orbit was a major problem in calculation of the moon's position. Newton had shown that the disturbing action of the sun would necessarily produce perturbations of the same general character as those recognised, and in the case of the motion of the moon's nodes and of her apogee, he was able to get a fairly accurate numerical result.[39]

The lectures discuss the cause of the tides, the precession of the equinoxes and the argument regarding the parallax of the fixed stars. Gregory maintained that Flamsteed's claim to have demonstrated the parallax of the fixed stars rested on a very weak foundation, but Whiston does not agree.

> "... it is most certain that this evasion of Dr. Gregory's, whereby he would show that the annual motion of the earth doth not follow upon Mr. Flamsteed's observations is not small error of his and leaves a blemish upon a work otherwise valuable for demonstrations strictly geometrical." (40)

Further lectures deal with attractive forces between bodies and show that for a homogeneous sphere the total mass can be considered as acting at the centre of the sphere. The laws of reflection and refraction of light as applied to plane and spherical surfaces are explained.

Pressure in fluids is discussed in Lecture XXVII. Whiston, like Newton, expresses his belief in the physical reality of particles which constitute an elastic fluid and which repel each other.*

> "If the density of a fluid, composed of particles which do flee from each other, be as the compression; so that if the pressing force be two or four or eight fold, the density thence arising is so likewise."

and in the same Lecture XXVII,

> "Since therefore it is manifest by experiments that the density of our air, compressed and rarified by turns is proportional everywhere to the compressing force, or the compression itself; it seems that the Air consists of particles whiche flee from or chase away one another in inverse proportion of the distances."(41)

Whiston discusses the physical basis for the propagation of sound and of light, and concludes that the rectilinear propagation of light can be explained by an actual flow of real corpuscles from the source. Newton never committed himself in writing to such a clear physical explanation of the phenomenon.

Lectures XXXII to XXXVII set out,

> "the Nature of things and the philosophical causes of these phenomena both astronomical and physical and the true system of the world; and that we set before you the frame and constitution of the same system as far as it depends on the principles already laid down." (42)

---

* By Dalton's time, it was a dogma that Newton had proved all gases to consist of self-repellent atoms. Perhaps Whiston started this interpretation of Newton's ideas. For a further discussion of this point, see <u>John Dalton and the Progress of Science</u>, Manchester University Press (1968), Editor D.S.L. Cardwell.

These six lectures demonstrate that the six planets and their satellites obey all the laws already outlined in the preceding thirty-one lectures.

In Lecture XXXII, Whiston refers to Kepler as the parent of Newtonian philosophy, an indirect criticism of Newton's lack of appreciation of Kepler's achievement.

By 1716 when the first English Edition of Whiston's lectures was published, Newton's second edition of the Principia (1713) was available, but Whiston did not revise his lectures to any considerable extent either for the edition of 1716 or the second Latin edition in 1726.

Lecture XXXVIII on Comets is drawn largely from Newton's observations on Comets in Book III of the Principia.

The final section consists of the text of, and Whiston's commentary on, A Synopsis of the Astronomy of Comets, by Edmund Halley. The great comet of 1680 was visible for four months continuously and the Royal Observatories at Greenwich and Paris had excellent opportunities for studying its motion. The most important observations were made by Cassini and Flamsteed. Newton proved that the comet of 1680 moved round the sun in a parabolic orbit and described areas taken at the centre of the sun proportional to the times. Halley tabulated all the known data on comets from 1337 to 1698, and constructed tables for calculating the motions of comets in parabolic orbits.

Halley at first implies that most comets move in parabolic orbits and therefore visit the solar system on one occasion only. However on considering the frequency of the appearance of comets he remarks,

> "tis highly probable they move in very eccentric
> elliptical orbits and make their returns after
> long periods of time ... And indeed there are
> many things which make me believe that the comet
> which Apian observed in the year 1531 was the
> same which Kepler and Longomontanus more
> accurately described in the year 1607 and which
> I myself have seen return and observed in the
> year 1682." (43)

This comet of 1682 has been named Halley's Comet and has a periodic time of 75-76 years. Its last appearance was in 1910.

Nicolas Fatio de Duillier proposed that the rare phenomenon of the transit of a comet would provide an excellent opportunity for the determination of the earth-sun distance.

In his commentary on Halley's _Cometography_ Whiston carefully allocates the praise to Kepler and Newton who were both concerned in the elucidation of comet theory.

> "... the illustrious Newton, writing his Mathe-
> matical Principles of Natural Philosophy, demon-
> strated not only what Kepler had found, did
> necessarily obtain in the planetary system; but
> also that all the phenomena of Comets would
> naturally follow from the same principles." (44)

Whiston's two books of lectures, _Astronomical Lectures_ and _Sir Isaac Newton's Mathematic Philosophy more easily demonstrated_ were adequately illustrated and were in use as textbooks for undergraduates for at least fifty years after their publication.*

---

* I. Bernard Cohen, in the Johnson Reprint Corporation
edition (1972) of Whiston's _Sir Isaac Newton's Mathe-_
matical Philosophy more easily Demonstrated, again
renders a double service to scholars in providing a list
of titles for the 41 lectures and in the preparation of
an index of persons mentioned in the text.

In May 1751, Ezra Stiles (1727-1795) wrote from Yale College, New Haven, Connecticut, asking Whiston for further copies of his scientific works for the library.

> "We have in our Library your philosophical and astronomical lectures which by constant perusal and study are almost worn out ... we should esteem it an honour if you would please to favour us with a copy of your works, especially Astronomy & Philosophy or any part of them as your benevolence may direct." (45)

4.4   EXPERIMENTAL   LECTURE   COURSES   FOR   THE   PUBLIC

In his Memoirs, Whiston recalls that in 1707 he and Roger Cotes began their first course of Philosophical Experiments at Cambridge.

> "In the performance of which, certain hydrostatic and pneumatic lectures were composed; they were in number twenty-four; the one half by Mr. Cotes and the other half by myself. Which lectures were also afterward made use of in the like courses which Mr. Hauksbee and I performed many years in London." (46)

At the same time, John Keill (1671-1721), at Oxford, was giving a course of lectures very similar to those of Whiston. Keill had attended school at Edinburgh and later attained distinction at Edinburgh University in mathematics and philosophy under David Gregory. About 1710, Keill gave the first course of lectures on Newtonian experimental philosophy at Oxford. His lectures <u>Introductio ad Veram Physicam</u> (1701) and <u>Introductio ad Veram Astronomiam</u> (1718) became standard textbooks and the former was well known on the continent.

John Theophilus Desaguliers (1683-1744) was a pupil

of Keill at Oxford and from 1710 was engaged in lecture
courses on experimental philosophy first at Hart Hall and
later in London.

One of the earliest courses of experimental philosophy
was given by James Hodgson (1672-1755) between 1704-1706 at
the request of Francis Hauksbee, senior, who acquired a
reputation as an experimentalist.*   His book, Physico-
Mechanical Experiments on Various Subjects (1709) was later
translated into French and Italian.   In the preface to this
work, Hauksbee states his belief in the importance of
experimental science.

> "The learned world is now almost generally convinced
> that instead of amusing themselves with vain
> hypotheses which seem to differ so little from
> romances, there is no way of improving Natural
> Philosophy but by demonstrations and conclusions
> founded upon experiments judiciously made."   (47)

Whiston and Humphrey Ditton (1675-1715) succeeded
Hodgson, and lectured alternately in experimental philosophy
at Hauksbee's house in Fleet Street, London.   Ditton had
already published in 1705 a short exposition of the Fundamental Theorems of Newton's Principia.

> "In this little treatise I have endeavoured to
> explain that part of Mr. Newton's Principia
> which is fundamental to the rest of the devine
> book; and particularly to the system of the
> world.   And I have taken care to render things
> so far plain and easy that they may be understood
> by all those that will bestow some application
> upon them and have a competent share of the
> necessary furniture, namely Algebra and Conic
> Sections."   (49)

---

*   Z.C. von Uffenbach recalls his visit to Hauksbee
    (senior) in 1710 and saw experiments with pumps and
    optical demonstrations.   (48)

After the death of Hauksbee in 1713, Whiston gave a series of lectures in co-operation with Francis Hauksbee (junior) (1687-1763), in which Hauksbee performed the experiments and Whiston read the explanatory lectures.

Whiston's A Course of Mechanical, Magnetical, Optical, Hydrostatical and Pneumatical Experiments was published in 1714 probably by John Senex.

A programme of the experimental work to be covered in a course of 28 days is outlined under seven headings:

Mechanics - 7 lectures.
The subject matter included Newton's three laws of motion, levers, pulleys, machines, pendulums, elastic and inelastic collisions.

Magnetics - 2 lectures.
Loadstones, location of magnetic poles using iron filings, laws of attraction and repulsion, spherical loadstones or terrellae, dipping needles, determination of longitude.

Optics - 6 lectures.
Laws of reflection at plane and concave surfaces, Newton's reflecting telescope, refraction, the eye, optical instruments including microscopes and telescopes, refraction due to atmospheric effects, rainbows, colour experiments.

Hydrostatics - 3 lectures.
Flotation, torricellian tubes, pumps, syringes, pressure in fluids, measurement of specific gravity.

Pneumatics

Illustrated by experiments for the most part tubular, being such as were wont to be made before the air-pump was invented - 3 lectures.

Barometers, thermometers, hygroscopes.   The density of the
air, the limits and state of the atmosphere, syringes,
pumps, syphons, artificial lungs.

The more known properties of the air established by the air
pump and other engines - 4 lectures.
The air pump, instruments for condensation and transfer of
air, gauges, measurement of the specific gravity of air,
air pressure demonstrated by bladders, fountains, the diving
bell; the boiling of liquids in vacuo.

The more hidden properties of the Air considered by the help
of the like engines - 3 lectures.
The influence of air as to the cause of magnetism, the
sphericity of drops of fluids; air and its influence on
sound, fire and flame, air and animals; phosphorus in vacuo,
new experiments with mercurial phosphori and experiments on
static electricity.

On closer examination this course of experiments given
by Whiston and Hauksbee (junior) seems to have a double
origin.   The sections on Mechanics, Magnetics and Optics
are probably Whiston's own contribution; the remaining
sections on hydrostatics and pneumatics are the work of
Francis Hauksbee (senior), Physico-Mechanical Experiments
(1709). In some cases the experiments are identical and
the engravings of apparatus in both works bear close
resemblance.

This publication of 1714 did not include the text of
the lectures but was in the form of an illustrated laboratory
notebook with 20 plates of engravings of apparatus.   In
some cases, 6 or more diagrams appear on a plate and the

opposite page describes the apparatus and explains its use.
Blank pages are interleaved with the plates indicating that
perhaps it was intended for use as a laboratory notebook.
Fourteen of the engraved plates are by Sutton Nicholls, the
remaining six by John Senex.   A second edition appeared in
the same year (1714).

The subscription for Whiston's course of lectures with
Hauksbee was three guineas; one guinea to be paid at the
time of subscription and the remainder on the first day of
the course.   Subscriptions were accepted at Whiston's
house in Great Russell Street and also at Mr. Hauksbee's
residence in Crane Court, Fleet Street, where the course
was planned to take place.*

Whiston's lecture courses and occasional astronomical
lectures given in the London coffee houses were no doubt
a source of financial support during the years after he left
Cambridge.   He was fortunate in having Joseph Addison and
Richard Steele as his patrons at this time.   Joseph Addison,
poet and statesman, was secretary to the Lord Lieutenant of
Ireland in 1709.   Richard Steele was an essayist, dramatist
and politician and in 1709 assumed the editorship of the
Tatler.   The "wits" of the day held their meetings in coffee
houses such as Will's and Button's.   Whiston tells how
these two friends came to his assistance,

---

* Marjorie Nicolson and G.S. Rousseau give full coverage
  to the impact of Whiston's astronomical and experimental
  lectures in their publication, This Long Disease,
  My Life (Alexander Pope and the Sciences), Princeton
  University Press (1968).

"Mr. Addison ... with his friend Richard Steele
brought me upon my banishment from Cambridge
to have many astronomical lectures at Mr.
Button's Coffee House near Covent Garden, to the
agreeable entertainment of a good number of
curious persons and the procuring me and my
family some comfortable support under my
Banishment."   (50)

Whiston's course of lectures with Hauksbee was advertised in The Englishman of Jan. 11, 1714.   This particular
set of lectures seems to have started, not in Fleet Street
but at Button's Coffee House.   The Englishman of Jan. 26th
1714 reports that:

"Mr. Whiston's Mathematic lectures will be this
day removed from Mr. Button's Coffee House to
a larger room close by at Mr. Dale's an upholst-
erer over the corner of the nearest piazza in
Covent Garden."   (51)

This seems to indicate that the lecture course was
successful.

Whiston's astronomical lectures were either organised
into courses covering a wide range of astronomical topics
or centred upon particular phenomena such as the solar
eclipse of 1715.   As early as 1713, Alexander Pope
attended what must have been one of Whiston's astronomical
lectures, and wrote to John Caryll on Aug. 14th, 1713:

"You can't wonder my thoughts are scarce consistent, when I tell you how they are dis-
tracted ...   This minute, perhaps, I am above
the stars, with a thousand systems about me,
looking forward into the vast abyss of eternity,
and losing my whole comprehension in the bound-
less spaces of the extended creation, in
dialogues with Whiston and the astronomers."   (52)

The younger Hauksbee and Whiston circulated a leaflet
advertising An Experimental Course of Astronomy possibly in

1714 or 1715.*

The course was to place special emphasis on the use of instruments and observation of celestial phenomena.

> "The telescopes of the common sort; and reflecting telescopes, quadrants, micrometers, and other astronomical instruments used in this course will be shewn and explained as they come to be applied in practice." (54)

This suggests that Whiston gave the lectures while Hauksbee demonstrated the instruments. In 1714 Hauksbee was already selling astronomical instruments as well as pneumatic and electrostatic apparatus. An advertisement for the apparatus made and sold by Francis Hauksbee was printed.

> "Air-pumps or engines for exhausting the air from proper vessels, with all their appurtenances; whereby the various properties and uses of that fluid are discovered and demonstrated by undeniable experiments. Engines for the compression of air: Fountains in which the water,

---

* The proposed course of 25 lectures was concerned with a study of the solar system and the determination of such items as the true meridian, measurement of the diameters of the sun and planets - the phenomena of tides.

The B.M. copy of this leaflet is provisionally dated as 1730. Two further items throw this date in doubt. Manuscript alterations to the sheet indicate the reduction of the subscriber's fee from five guineas to three guineas and also announce that the lecture course is due to begin on Monday February 22nd at six in the evening. According to the Universal Calendar Feb. 22nd fell on a Monday in 1714 and 1720.

The manner of computing lunar and solar eclipses was to be demonstrated using Whiston's Copernicus. The two famous total eclipses of the sun of April 22nd 1715 and May 11th 1724 were to be shown as examples. The inclusion of the solar eclipse of 1724 does not necessarily indicate that the date of the lecture course was later than that year. As early as 1715 Whiston had prepared An Account of the Eclipses of the Sun 1715 and 1724, engraved on copper by John Senex. (53)

or other liquor, is made to ascend by the force
of the air's spring.   Syringes and blow-pipes
with valves for anatomical injections.   Hydrostatical balances for determining the specific
gravity of fluids and solids.   The engine and
glasses for the new way of cupping without fire.
Scarificators which at once make 10, 13 or 16
incisions.   Weather glasses of all sorts, as
barometers, thermometers.   Reflecting telescopes
by which in so short a length as six feet, all
that has hitherto been discovered in the heavens
(by the longest telescope of the common construction) may be observed.

All the above mentioned instruments, according
to their latest and best improvements, are made
and sold by Francis Hauksbee in Crane Court,
Fleet St. London."   (55)

There was also to be a demonstration of the calcula-

tion of the velocity of light from the eclipses of Jupiter's

satellites (Lecture XVII).   Parker's Ephemeris of the

Celestial Motions was to be explained.   George Parker

(1651-1743) commenced the publication of an almanac with the

title Mercurius Anglicanus in 1690.   In 1703 it was called

A Double Ephemeris and from 1707 Parker's Ephemeris.(56)

Whiston probably repeated his lecture courses at

intervals.   There is evidence that in February 1720, he

had begun a further set of astronomical lectures.   William

Whiston (junior) wrote to his sister Sarah,

Feb. 11th 1720.

"... My father has begun his course of Astronomy
with about 21 subscribers."

The subscription fee was probably three guineas and sixty

three guineas would meet his household expenses for several

months.

The total solar eclipse of 22nd April 1715 had been

foretold by Flamsteed, Halley and Whiston.   Both Halley and

Whiston had schemes printed, giving full information about
the eclipse, for sale to the public.   There was widespread
interest in the forthcoming eclipse, since, according to
Halley's calculations, no total eclipse of the sun had been
visible from England since 1140.   Whiston gave public
lectures on the subject and must have sold many schemes.[57]

> "This most eminent eclipse, 1715, was exactly
> foretold by Mr. Flamsteed, Dr. Halley and
> myself ... I myself by my lectures before, by
> the sale of my schemes before and after; by the
> generous presents of my numerous and noble
> audience ... gained in all £120 by it.   Which
> in the circumstances I then was, and have since
> been, destitute of all preferment, was a very
> seasonable and plentiful supply: and as I
> reckoned, maintained me and my family for a whole
> year together." (58)

The Daily Courant of March 6th 1716 advertised a
public lecture to be given by Whiston on March 16th 1716 on
the Surprising Appearances in the Air on Tuesday 6th Instant.
This lecture was given in the Censorium, a hall in York
Buildings, which according to Richard Steele's plan was used
as a little theatre and lecture hall.[59]

Whiston was still engaged in courses of Astronomical
lectures in 1736.   He had foretold the transit of Mercury
on October 31st 1736.   In his publication The Astronomical
Year 1736 he records,

> "With the explication of which [the transit of
> Mercury], and of the great use such transits
> may sometimes be, in order to discover the
> parallax and distance of the sun, I began my
> astronomical lectures last winter." (60)

Whiston would no doubt have liked to introduce into
his public lectures on natural philosophy references to his
interest in Primitive Christianity and other theological

notions relating to his heretical ideas, but probably he
was persuaded to confine himself to scientific matters.
Henry Newman, American secretary to the Society for Promoting
Christian Knowledge warned Richard Steele of Whiston's
tendencies:

> "I only beg leave to suggest one thing to you,
> that when he [Whiston] does [lecture] ... that
> you will be pleased to conjure him silence upon
> topics foreign to the mathematics his Conversa-
> tion or lectures at the Coffee houses. He has
> an itch to be venting his notions about Baptism
> and the Arian doctrine but your Authority can
> restrain him at least while he is under your
> guardianship." (61)

Courses of experimental lectures such as those given
by Whiston and Hauksbee became very popular. In 1722 a
similar course was advertised by Benjamin Worster:

> "A Course of Experimental Philosophy to be per-
> formed by Benjamin Worster and Thomas Watts at
> the Academy or Accomptant's office for qualify-
> ing young gentlemen for business in Little Tower
> Street." (62)

Worster's syllabus was almost identical with Whiston's
course of 1714.

3.5   THE COPERNICUS

Whiston was present at the meeting of the Royal Society
on February 10th 1715. The Journal Book Copy for that
meeting twice makes mention of Whiston:(63)

> "Mr. Whiston presented the Society with a book of
> his lectures in Experimental Philosophy for which
> he had the thanks of the Society."

This probably refers to the English translation of his
Astronomical Lectures (1715). A later entry for the same

meeting records,

> "Mr. Whiston showed the Society an instrument he
> had contrived for finding out and constructing
> both solar and lunar eclipses and he desired
> the opinion of the Society concerning it.   It
> was ordered that Dr. Halley be desired when he
> comes to town to view it."

This instrument was Whiston's <u>Copernicus</u>.   In the same year,

Whiston published a small duodecimo manual, <u>The Copernicus</u>

<u>Explained</u> (1715) describing the construction and use of the

instrument.

> "This astronomical instrument (which is made
> agreeably to the Copernican system and therefore
> by me named the Copernicus) consists (besides
> the immovable circle on the outside, and the
> small moveable central circle within) of ten
> intermediate concentrical annuli, or broad
> circular Rings, fitted to revolve one within
> another;   but so, as to be capable of being
> fixed by small pins in any situation whatsoever.
> Six of the circles towards the centre are so
> contrived, that they may be taken away upon
> occasion;   and yet when they are put in they are
> fast connected to the frame, as well as the other
> and revolve as freely as they do.   There is also
> a terrestrial globe, of nine inches diameter,
> placed under the inner circles, with its hour
> circle, turning along with it in its diurnal
> motion: and when those circles are removed, the
> globe may be so elevated and fixed at any height,
> and so regulated by screws as to be ready for the
> exhibition of those eclipses, which the six outward
> circles assist us to discover;   to which last does
> also belong a rule with a groove, containing an
> angle of $5°37'$ for the moon's path in eclipses.
> There is also a round plate of glass, with 12
> concentrical circles therein, for the 12 digits in
> solar eclipses;   whose diameter bears the same
> proportion to the diameter of the globe that the
> apparent semidiameters of the sun and moon put
> together, do really bear to the disc of the earth
> in those eclipses.   There is besides a map of
> the moon, with 6 concentrical circles, for the 12
> digits in lunar eclipse ...   There is also a
> dark circle representing that section of the earth's
> conical shadow, along which the moon passes in its
> own eclipses;   and is so much less in proportion
> than the diameter of the globe, as is that of the
> real corresponding circle to that of the earth
> itself."   (64)

After this description, Whiston gives nineteen problems which the Copernicus may be used to solve; for example:

Problem VI   To find the Sun's True Place in the Ecliptic for any time past present or to come.

Problem IX   To find whether there will be either a solar or lunar eclipse, at any conjunction or opposition, for any time past, present, or to come.

Problem XII   In a solar eclipse, considered in particular, and with regard to any single place, to find when it did, or will begin and end; whether it was or will be there total or partial; and if partial how many digits were or will be eclipses: with the other particular circumstances of the same in that place.

Frequent reference is made to the forthcoming total solar eclipse of April 22nd 1715.

The Appendix consists of three sets of tables which are necessary to make use of the instrument,

1.   A Table of the Mean Place of the Moon's Apogee and Node and of the Moon itself, in the beginning of the several centuries before and since the Christian era. The Table covers the period 1000 B.C. to 1701 A.D.

2.   A Table of the General Eclipses of the Sun within less than a Day under or over till A.D. 1754.

3.   A Table of the Eclipse of the Moon within less than a Day under or over till A.D. 1754.

The instrument, widely advertised, was made by John Senex at the Globe near Salisbury Court in Fleet Street and cost six guineas. Whiston also sold the instrument at his own house in Cross Street, Hatton Gardens. Enquiries have not brought to light a surviving model so it is only possible to conjecture its appearance from the description given in The Copernicus Explained. Ita arrival on the market was timely in that, as already mentioned, there was widespread interest among the public in the forthcoming eclipse. Whiston's Copernicus appeared at least a month before the event and he used it to demonstrate the calculation of eclipses in his public lectures.

It is clear from Whiston's own remarks that he intended the Copernicus to be a tool in his chronological studies.

> "I intended the Copernicus ... for the examination of all the ancient eclipses, that could possibly be seen in any parts of the world, of which we have any ancient histories preserved ... that so no historians or chronologers might ever be at a loss hereafter for the circumstances of such eclipses as are mentioned by any ancient author ... Accordingly I calculated by it the eclipses of the sun and moon for four several periods of eclipses ... A table of which eclipses, 250 in number, I have by me now." (65)

The following advertisement for Whiston's Copernicus appeared in 1715 at the end of the second edition of A New Method for Discovering the Longitude both at Sea and Land (1715). The same advertisement mentions proposals for a new set of maps which were to be made as a result of a new survey. There is no evidence that these maps were ever made or sold.

"Advertisement

> Lately published, proposals for a new survey of England and Wales, according to this method for discovering the Longitude; and for a new sett of correct maps of the several counties according to such a survey, by subscription at two guineas the whole sett, bound in pastboard, are to be had gratis at Mr. Whiston's in Cross Street, Hatton Gardens, at Mrs. Ditton's in Christ's Hospital; and at Mr. Senex's at the Globe near Salisbury Court, Fleet Street; and at the booksellers of London and Westminster, and of several county towns."

"As also Mr. Whiston's Copernicus, or
Universal Astronomical instrument for finding the
new and full moons and the eclipses of the sun and
moon; with the exhibition of them both by an easy
and natural method, never before published;
together with the places and aspects, both helio-
centrical and geocentrical, the directions, stations
and retrogradations of the planets, both primary
and secondary, for any time past, present, or to
come; according to the true or Copernican system:
with a small book, giving plain directions for the
use of it. The Prices are four guineas with the
globe and frame; three with only the frame; and
one and a half without either." (66)

In 1724, Whiston published a proposal to make an

orrery and an improved version of the Copernicus, but there

is no evidence that these instruments were in fact made.

"Advertisement

Mr. Whiston proposes to make an Orrery for the
explication of the motions of all the planets
and comets; and a Copernicus Improved, for the
easy discovery of all solar eclipses whatsoever,
from the beginning of Astronomy till A.D. 2000
and for the natural and universal exhibition of
the same to the eye in a few minutes time, and
this with a globe of thirty inches diameter.
Price of both 20 guineas." (67)

4.6   RECORDS OF VARIOUS ASTRONOMICAL PHENOMENA

Whiston took an active interest in the astronomical

phenomena of his time. This interest was not confined to

astronomical events which could be predicted, such as solar

and lunar eclipses, but included cataloguing and analysing

unexpected phenomena such as the appearance of comets,

meteors, mock-suns or auroral displays.

Whiston was particularly well equipped to bring all

these phenomena to the public notice and thus foster the

current interest in astronomy. He used three channels of

communication, the public lecture, short tracts or pamphlets, and charts or engravings which were often very elaborate in their decoration and seemed to find a ready market. Whiston's efforts to supply the public with information on these matters provided him with a substantial part of his income.

Solar Eclipse of 1715

Flamsteed wrote to Whiston in November 1714 referring to the calculation of the eclipse.

> "The Observatory.
> November 30th 1714

> "I have yours of the 15th last which brought
> the Calculation of the ... eclipses for the
> next year 1715. I thank thee heartily for
> them." (68)

Richard Allin, Whiston's friend in Sidney Sussex College, Cambridge, wrote to him two days after the great eclipse.

> "April 24th 1715. Cambridge.

> "I had no instruments to make any accurate obser-
> vation of the eclipse but without doubt Mr. Cotes
> had done it very carefully ... Till the total
> obscuration was over, we had as fine and clear a
> morning as could be desired; but as we came out
> of the penumbra, the sun was sometimes clouded
> yet so as the clouds assisted us in viewing the
> several appearances of the eclipse with our naked
> eyes. Between 8 and 9 o'clock, I went with
> several of our College into the fields behind
> Barnwell, that I might have the opportunity of
> seeing your ball of fire, which (as an advertise-
> ment had said) was to be thrown up from Hampstead
> Heath at the middle of the total darkness; but
> none of us could see anything of it, nor did some
> others, who watched carefully for it on the Castle
> Hill, at all perceive it. As to the darkness, it
> was not so great as expected by some. We could
> very clearly see the hills at a distance and
> several trees and hedges some furlongs off and the

green corn nearer at hand, with birds flying
over it, as also the countenances and cloaks of
each in the company.   What light there was seems
derived from a bright ring which closely begirt
the body of the Moon, immediately after the last part
of the sun disappeared;   notwithstanding which ring,
it was so dark that I could not read in the small
Cambridge Virgil (printed A.D. 1702) which I had put
into my pocket, though another of the company
affirmed that he could.   I saw Jupiter and Venus
but could not see Mercury, tho' everyone besides
saw it clearly and said it was nearer to Venus ...
Some of our company, as also those on Castle Hill
saw one or two fixed stars;   but none of us saw
any streaks of red light or anything (except the
ring before mentioned, which was not red but white)
from when the lunar atmosphere might be concluded.
And quaere whether that ring did not arise, either
from the inflection of the sun's light as it
passed near the body of the moon (Princip. Math.
p. 207, edit 2nd) or more probably from an atmosphere about the Sun.   None of us could see anything like the sun's Milky Way ..."   (69)

This account of the eclipse of 1715 is probably one
of the most interesting unpublished observations of the
phenomenon.

Surprising Appearances seen in the Air 1716 and 1719

Whiston published two reports on atmospheric phenomena:
An Account of a Surprising Meteor seen in the Air, March the
6th 1716, at Night (1716) and An Account of a Surprising
Meteor seen in the Air, March 19th 1719 at Night.

The term meteor was commonly used to cover many
classes of atmospheric phenomena.   Reports were read and
written of airy meteors, aqueous meteors, luminous meteors
and igneous or fiery meteors.

The event which occurred on March 6th 1716 was a
magnificent auroral display.   Whiston witnessed this event
from London and his 1716 publication gives a first-hand
account of what must have been a wonderful scene.   The

display lasted from about seven o'clock until after midnight. Whiston gives thirteen extracts from historical
sources which describe former observations of the aurora
borealis, and eleven extracts of letters from observers in
different parts of the country, including places as far
north as Edinburgh and as far south as Lewes in Sussex.
Many different names are mentioned for the phenomenon,
Northern Twilight, Fighting Armies; Whiston himself seems
to be the originator of the term Northern Lights.* He
concluded that the exhalations of the Northern Twilight
are situated:

> "between the lower vapours which cause winds and
> storms and those higher exhalations which cause
> thunder and lightning ... These streams or
> pillars seem to me to have arisen from such
> violent fermentation and rarefaction in some
> parts of this fund of exhalations ... Nor is
> it strange that they went up perpendicularly,
> since tis the greater gravitation of the air
> which in part at least must cause and govern
> that ascent." (71)

> "(21) Since the usual direction of a line drawn
> through the middle of the vapours that cause
> these appearances is certainly with us about 14
> or 15 degrees from the North to the West, or
> (67°) the same that every magnetic needle lies
> along; and that the centre of that cupola which
> they sometimes represent is nearly on the same
> angle, that a magnetic dipping needle, vibrating
> in the plain of its magnetic meridian, lies in,
> Dr. Halley's notion that the effluvia which
> occasion them are magnetic effluvia seems
> exceedingly probable."

A further note adds,

---

* The British Museum copy of the second edition of
  Whiston's <u>Account of a surprising Meteor seen in the Air</u>
  1716 has interesting manuscript notes on page 67. These
  were written in 1736 and show that Whiston was aware of
  the connection between auroral displays and terrestrial
  magnetism. (70)

"N.B.   Whatever hypotheses have been made about
these meteors or northern lights, none pretend to
give any sure account why they have been so much
more frequent for 20 or 21 years together from
1715 till 1736 than formerly: which fact yet I
take to be undenyable."   (72)

Whiston did not personally observe the meteor of
March 19th 1719.   He begins his pamphlet with,

"a faithful relation of the phenomenon itself,
and that from the original accounts of some of
the most intelligent of those that saw it."

Extracts are given from eleven accounts.   The event was
evidently the approach of a spectacular meteor or meteorite.
There was not enough evidence to determine whether a
meteorite fell into the sea, and there was no report of a
body falling on land.

Whiston quotes nine earlier accounts of similar
meteors.   He disagrees with the account of John Wallis
where they are identified as comets.(73)   Halley had
expressed his views about the nature of such phenomena.(74)
He postulates,

"some collection of matter formed in the aether,
as it were by some fortuitous collection of
atoms; and that the earth met with it as it
passed along in its Orb."   (75)

Whiston disagrees with Halley as he cannot allow the
existence of a fortuitous collection of atoms not subject
to gravitation.   His own solution is,

"That they are no other than prodigous storms
or blasts of thunder and lightning in the upper
regions of our air."   (76)

This work includes a diagram of the track of the meteor drawn
by Nicolas Fatio de Duillier who had observed the meteor at

Worcester.

A second edition of this work was printed in 1719 to which Whiston added a postscript dated May 25th 1719, entitled Mr. Whiston's Vindication of his Account of the late Meteor, from the different account given of it by Dr. Halley in Philosophical Transactions No. 360.

Halley calculates the meteor to have made its appearance at a height of 60 or 70 statute miles above the earth. Whiston objects that Halley has already proved that the height of the atmosphere does not exceed 50 statute miles. Whiston is unwilling to admit that the meteor could have been visible above the limit of the atmosphere and he claims that other observations agree with a lower altitude.

"I believe this meteor was never beyond the limits of our sensible atmosphere, because the distances where it has been seen better agree to this lower altitude, than to that assigned by Dr. Halley. Thus we hear no further accounts of it than Drogheda in Ireland, Peterhead in Scotland; Cleves in Germany; Paris in France; and such neighbouring places. Whereas had it been 70 measured miles high, it would have been further heard of." (77)

Whiston, relying on the observations of Fatio de Duillier, calculates the height of the meteor to be 47 geographical miles.

Appearance of Mock-Suns, 1721

Whiston communicated to the Royal Society An Account of two Mock-Suns and an Arc of a Rainbow inverted with a Halo, and its bright arc, seen on Sunday Oct. 22nd and 23rd 1721 at Lyndon, Rutland. (78)

This account was read to the Society at its meeting on Wed. 26th October 1721. The phenomenon followed a display

of the Aurora Borealis on the previous night.   Whiston
described it in some detail, with diagrams.

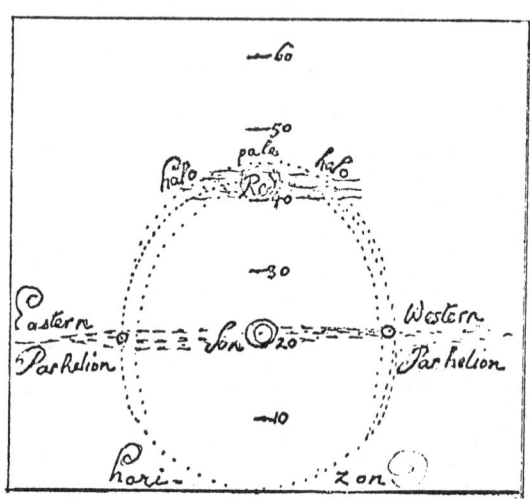

"The mock suns were evidently red towards the
sun, but pale or whitish at their opposite
sides - as was the halo also.   Upon casting
our eyes upwards, we saw an arc of a curious
inverted rainbow about the middle of the
distance between the top of the halo and our
vertex."

The same phenomenon appeared the next day Monday,
Oct. 23rd near noon and each day lasted 1½ to 2 hours.
Whiston reports the appearance of a halo on Thursday Oct.
26th about 9 o'clock.

"...   As I was coming in the Northampton Coach
towards London, about 9 o'clock, the halo
returned larger and clearer than before;   and
the two mock-suns just attempted an appearance
therein, as on Sunday but the air becoming
thicker and thicker towards rain, I saw them no
more."

Halley reported on the same phenomenon to the Royal
Society on October 16th 1721.

"Dr. Halley read an account of what he observed upon the appearance of a crown or circle about the sun which happened this morning about half an hour past ten." (79)

Whiston noted a magnificent auroral display on September 11th 1721. He correlated his observations of this event with friends in Bath and Northamptonshire.

"We had there [Rutland] Sept. 11th in the evening the lightest and most remarkable Aurora Borealis with its unaccountable motions and removals that I ever saw, excepting that original one March 6th 1716 - that was seen in Northamptonshire and Bath and elsewhere, that the vertex of the columns which shot upwards, was not our vertex but evidently 15 or 20 degrees towards the south and that the wind was in Rutland North, as I observed myself, at Bath, West; as Mr. Molyneux observed; as I am informed by Robert Clarke in Northamptonshire South, all at the same time, - which deserves particular reflection." (80)

He knows the explanations of haloes, mock-suns and inverted arcs of rainbows as given in the posthumous works of Huygens.*

Whiston's observations of the mock-suns of Oct. 22nd were printed in the Philosophical Transactions Abridged (1733), by Reid and Gray among many other such observations.

Solar Eclipse of 1724

In 1724, Whiston published The Calculation of Solar Eclipses without Parallaxes (94 pp.). This included a

---

* Christiani Hugenii,
Opuscula Posthuma quae continent Dioptricam (1703)
Lugduni Batavorum
Dissertatio de Coronis et Parheliis (p.293-366)
including eight plates of figures.
Extracts from Huygens' works were printed in the Lowthrop
Abridgements of the Philosophical Transactions (1716)
Diagrams facing page 204, Vol. 2.

calculation for the total eclipse of the sun for May 11th 1724. The same work included two additions which are discussed in Chapter 3, namely, A Proposal how, with the latitude given, the geographical longitude of all the parts of the earth may be settled by the bare knowledge of the duration of solar eclipses, and especially of total darkness, and secondly, An account of some late observations made with Dipping Needles in order to discover the longitude and latitude at sea.

Whiston calculates the solar eclipse of 11th May 1724 for three places, Dublin, London and Paris; on comparing his results with those of Halley, he finds that there is not full agreement.

> "N.B.   Dr. Halley's times are still about 5' sooner than mine and his general duration about 7', and his duration at London about 1' longer.
>   I cannot tell the reason why my original calculation of this general eclipse which has been carefully made according to Sir Isaac Newton's Theory of the Moon does here so much differ from Dr. Halley's determination as 5' in time; especially since both these methods did very well agree in the past celebrated total eclipse April 22nd 1715."   (81)

Whiston was in contact with astronomers at European observatories concerning their observations of the eclipse of 11th May 1724 and on December 17th of that year he read a paper to the Royal Society communicating observations from Berlin, Nuremberg and Hamburg. (82)

Extraordinary 'sun-spot', September 1730

This was reported by Joseph Wasse (1672-1738) rector of Aynho, Northamptonshire, and Whiston communicated the account to the Royal Society.

> "... concerning an extraordinary spot seen by the
> naked eye in the sun on 23rd September about
> quarter of an hour before sunset ... several
> persons saw a large spot which by their account
> should be near the 20th part of the whole
> diameter ... it was perfectly round and black
> ... The horizon was made by an open fallow
> field wherein nothing intercepted their sight.
>   Mr. Wasse receiving the account viewed the
> sun the next day with a seventeen foot tube but
> found it entirely free of spots."   (83)

The account ends with Whiston's suggestion that it could

have been a comet passing across the face of the sun.

The Astronomical Year 1736

Whiston noted eighteen astronomical phenomena for the

year 1736 which made him regard this year as one of great

significance.   To commemorate these events he published a

pamphlet of 26 pages entitled, The Astronomical Year: or

an Account of the many remarkable celestial phenomena of

the great year 1736, particularly of the late comet which

was foretold by Isaac Newton and appeared at its conclusion

(1737).

Many of these events Whiston had foretold in his

earlier publications.   Newton seems to have predicted a

return of the comet of 1668 which Cassini believed to have

a period of 34 years.   Maraldi in Rome saw what could have

been the tail of this comet in 1702 and it was due to

reappear towards the end of 1736.

The most notable of these phenomena were as follows:

a) The total eclipse of the moon, March 15th 1736.

Whiston himself observed this event on a clear evening.

b) Total eclipse of the moon, 9th September 1736 lasting
for 4 hours 12 minutes.   Whiston saw this from
Rutland but did not watch during the entire time.

c) The transit of Mercury on 31st October 1736.   Whiston
remarks,

> "I had the satisfaction of seeing it myself
> in Fleetstreet, through a smoked glass and
> a good telescope."

d) An annular eclipse of the sun Feb. 18th 1737 which
Whiston evidently saw from London.

e) The reappearance of the comet of 1668 which was observed
for fifty days between Jan 29th 1737 and March 20th
1737.   Whiston remarks:

> "so far as we yet know, Sir Isaac Newton
> is the very first man, and this the very
> first instance, when the coming of a comet
> has been predicted beforehand, and has
> actually come according to that prediction,
> from the creation till this very day."   (84)

The confusion about dates is due to the fact that until
1st January 1752, the English legal or ecclesiastical year
was considered to begin on 25th of March.   Thus, Whiston's
publication, the Astronomical Year 1736 in fact extends from
25th March 1736 to 25th March 1737.

Some of these astronomical phenomena were not visible
in England – others were unspectacular events such as the
occultation of Aldebaran on the night of 24th March 1737,
but he sees them as,

> "marking out some great designs of providence."[85]

In particular, these events signalised,

"the end of all persecution in Great Britain"
and Whiston earnestly hoped that the events described in
his work,

> "may be among the rest of the astronomical
> phenomena thereto belonging, introductive of
> the old primitive Christian religion into
> Britain and other parts of Europe."   (86)

The Comet of 1742

A letter to the editor of the Gentleman's Magazine
from G. Smith of Boothby near Carlisle reports the presence
of a comet,

> "On the 20th [February] about 4 o'clock in the
> morning, I accidentally discovered a comet in or
> near the extremity of the tail of the Serpent.
> ...   I have no instrument to make proper obser-
> vations, I therefore desire you would consult
> some of the eminent astronomers in London."   (87)

Whiston and Thomas Wright replied to the query.   According
to Whiston,

> "The comet was seen in the west last Thursday
> ... about Wednesday, it was in the Milky Way
> near the tail of the Eagle ... it is going
> almost northwards at the rate of about six
> degrees a day."

Thomas Wright (of Durham) observed the comet with
the naked eye and gave a simple sketch showing its movement
against the background of the stars.

Selections from Correspondence

Whiston engaged in wide correspondence about astronomical events.   He was in touch with many astronomers in
academic circles such as Roger Cotes, Edmund Halley, John
Flamsteed and Isaac Newton, and was also familiar with

persons interested in spreading philosophical and astronomical ideas to a wider public, such as Humphrey Ditton,
Francis Hauksbee, John T. Desaguliers. He occasionally
presented papers to the Royal Society and attended meetings
and felt free in 1748 to approach Martin Folkes the
President advising him of the importance of observing the
transit of Venus in 1761.[88]

In May 1751, E. Stiles wrote from Yale College,
Connecticut asking Whiston's assistance,

> "[I] should myself esteem it a particular
> honour if you would favour me with a few
> plans, draughts, schemes of solar eclipses,
> transits of Mercury and Venus over the sun, or
> any representation of the celestial motions,
> phenomena of the planets, comets and especially
> the comet of 1758. And since the periods of
> this comet seem to have been very unequal, you
> would oblige me with your judgement when it will
> next be in its perihelion." (89)

This request seems to indicate that Whiston's charts and
schemes were widely recognised as a source of reliable
information and were used in instruction.

4.7   MAPS AND CHARTS OF ASTRONOMICAL INTEREST

Whiston's New Theory of the Earth (1696) had a frontispiece of the Systema Solare and seven illustrations at the
end of the book which were of considerable help in understanding the text.

His first separate astronomical chart appeared in
1712, entitled A Scheme of the Solar System. This item
is now rare and copies are known to exist in the Map Room
of the British Museum[90] and in the Library of Trinity

College, Dublin.　It cost 2s. 6d. and was engraved by
John Senex.　It seems likely that Whiston drew up this
chart for use at his public courses of astronomical lectures.
This fine chart engraved by John Senex measures approximately
60 cm. x 70 cm.　It is basically a scale drawing of the
orbits of the planets and has insets showing their relative
sizes.　The orbits of the 24 comets listed by Halley are
all drawn.　Marked on each orbit of the comets is the
following information:　date of appearance, inclination
of its orbit to the plane of the ecliptic, location and
distance at perihelion, periodic time.　Five inset circles
of script type describe the constitution of the sun and
planets and explain that a hot body like the sun maintains
its definite size and shape due to the pressure of a
surrounding atmosphere.　Six further blocks of script
around the diagram of the solar system give a general description of the sun, planets and comets.

　　Whiston speculates that comets are capabl
deluges and conflagrations to planets and so may be the
instruments of divine vengeance upon the wicked inhabitants
of worlds.　Comets seem to be "chaoses" or worlds in confusion but are capable of a change to a nearly circular
orbit when they could become places fit for habitation like
the planets.

　　Whiston gives 900,000 millions of miles as a possible
distance of the fixed stars from the sun, basing this calculation tentatively on an annual parallax of about 45"
which Hooke and Flamsteed think they have discovered.
Whiston allows himself, to speculate about the more distant

parts of the Universe.

> "... of such vast and numberless systems [of Fixed
> Stars] ..... we know very little. Only so much
> we know of the planetary and cometary world and
> of the probability of vastly more among the
> fixed stars, (to say nothing of the noblest or
> invisible parts of the creation, nor of the
> particular phenomena here below)."

Whiston published a joint scheme for the solar eclipses of 1715 and 1724; probably in 1715. I have not traced this publication. He produced a second scheme of the eclipse of 1715 entitled: A Calculation of the Great Eclipse of the Sun, April 22nd 1715 in the Morning from Mr. Flamsteed's Tables as corrected according to Sir Isaac Newton's Theory of the Moon in the Astronomical Lectures with its construction for London, Rome and Stockholm, by W. Whiston M.A.[91] This scheme measures 9" x 12" and is a chart showing the passage of the moon's shadow across the corresponding arc of the earth's surface. Insets give a calculation of the eclipse and an explanation of its construction.

In 1723 (reprinted 1736) Whiston published A Scheme of the Transits of Mercury and Venus over the sun for 250 years.[92] The first edition was brought out in anticipation of the transit of Mercury on October 29th 1723 and the second edition was preparatory to the transit of Mercury on October 31st 1736 which was visible in its entirety from the latitude of London. This chart is beautifully engraved and the inset script makes it intelligible to anyone with even a slight knowledge of astronomy.

Transits of Mercury and Venus provide opportunities

for exact calculations of the sun's parallax and hence the determination of the magnitudes and distances of the heavenly bodies.   Whiston notes the transit of Venus observed in Lancashire by Horrox in November 1639.   Halley had stressed the importance of careful observation of the next transit of Venus May 26th 1761.   Whiston makes the point that the transit of Mercury eight years earlier in 1753 would provide an excellent opportunity for determining the earth-sun distance with a high degree of accuracy.

In 1733 appeared Whiston's <u>Map of Europe</u> on which were drawn the courses of eight famous eclipses of the sun.[93] This map, 50 x 60 cm. approximately, was probably taken from John Senex's own Atlas.   It was evidently produced at this time to coincide with the total eclipse of the sun, May 2nd 1733 visible in the Scandinavian countries between the latitudes of Stockholm and Copenhagen.

The map showed the path of 5 earlier eclipses, those of 431 B.C., 1652, 1706, 1715 and 1724.   Two future eclipses marked on the chart were those of Feb. 18th 1737 and March 21st 1764.   It is characteristic of Whiston that he chose to include an eclipse from antiquity on this chart indicating his continuing interest in chronological studies.

4.8   ASTRONOMICAL PRINCIPLES OF RELIGION

In 1717 Whiston published his work, Astronomical
Principles of Religion, Natural and Revealed.*

This work seems to embody every aspect of Whiston's
thought and endeavour.   Here it is possible to see
Whiston as natural philosopher and mathematician, a devotee
of Newtonian ideas and anxious to bring this new understanding of the physical world to the notice of the public.
The title of the work indicates Whiston's firm belief that
a study of the wonders of the physical universe undertaken
in the right frame of mind can only serve to engender a
greater reverence for the goodness of God the Creator.
It follows in the tradition established by John Ray,
The Wisdom of God in the Works of Creation (1691).

It is dedicated to Newton as President of the Royal
Society and to the Council and Fellows of the Society.
This dedication is significant in that Whiston was anxious
that admirers of Newton should be moved to arrive at the
same conclusion regarding the adoration of God through
His created works.

In the second (1713) and third (1726) editions of
Principia, Newton adds his famous Scholium Generale to
Book III, the System of the World.   In this Scholium Newton

---

*   A second edition of this work appeared in 1725.   The
     British Museum copy of this work (Shelf Number 874.1.5)
     contains manuscript notes and corrections by the author.
     Some of these corrections were made after 1745 as Whiston
     refers to his Sacred History of the Old and New Testament
     which was published in 1745.   He revises some of the
     astronomical data given in the early parts of the book
     in the light of discoveries.

expressed the sentiment that the contemplation of the harmony of the physical universe would lead man to adore and reverence the Maker and Lord of the Universe.

Astronomical Principles of Religion opens with a lengthy preface describing the frame of mind in which one should search for truth. The book is preceded by Milton's Hymn to the Creator and closes with Sir Richard Blackmore's poem of the same title.

It is divided into nine sections, the first five of which form an attractive exposition of contemporary astronomical ideas. The text includes eleven carefully executed engravings of astronomical interest by John Senex. These five sections give descriptive and factual information about the physical universe and Whiston makes many references to his own work, Sir Isaac Newton's Mathematical Philosophy more easily Demonstrated (1716) (English) and to Newton's own Mathematical Principles of Natural Philosophy (1687).

In Part V entitled Probable Conjectures as to the Nature and Uses of all the Parts of the System of the visible world, Whiston presents novel ideas about the possible use of the upper regions of the atmosphere and of the interior regions of the earth.

He seems to suggest the supernatural spirits, e.g. - the angels and spirits of the just inhabit the higher regions of the atmosphere.

> "The air expanded about the several planets, which, as to their elastical parts, are corporeal, but invisible, appear to be proper places for the habitation of not wholly incorporeal but invisible beings." (94)

Whiston revives the idea put forward by Halley that some of the heavenly bodies have central cavities,

> "that may be so fitted by Providence, as to afford
> Habitations to some creatures."

Whiston seems to have no doubt that the earth has such a cavity.

> "that the earth in particular has such a cavity
> seems clear from scripture, as well as it may
> be conjectured from Astronomy.   For when many,
> at least, of the souls departed out of the
> world, are there represented as gone down into
> the invisible world."   (95)

In Part VI <u>Important Principles of Natural Religion demonstrated from the foregoing observations</u>, Whiston concludes that from this true system of the world we learn that the Supreme God, the maker and governor of all things, is not a blind fate, or series of necessary causes and effects, but is a spiritual, a living and an active being, perpetually exerting His Divine Perfections in the whole universe.

Part VII is entitled <u>Important Principles of Divine Revelation confirmed from the foregoing Principles and Conjectures</u>.   Whiston reasserts the truth of the Mosaic account of creation and asserts that natural science and sacred chronology agree in dating the earth at not more than 7000 years old.   He condemns Halley's speculations that a measure of the increasing salinity of the oceans might lead to a much more extended time-scale for earth history.(96)

In Part VIII <u>Some Inferences shewn to be the Common Voice of Nature and Reason from the Testimonies of the most</u>

considerable persons in all ages, Whiston quotes extensively
from scripture and then makes frequent reference to other
more recent work in the tradition of natural religion.   He
refers to Ralph Cudworth's Intellectual System of the
Universe, John Ray's The Wisdom of God in the Works of
Creation, William Derham's Astrotheology, Robert Boyle's
Of the Veneration Man's Intellect owes to God, Molyneux's
Dioptricks and of course Newton's Philosophia Naturalis
Principia Mathematica.

The theme of Whiston's Astronomical Principles of
Religion could be summed up in Newton's words,

> "This most excellently contrived system of the
> sun, the planets and the comets could not have
> its origin from any other than from the wise
> conduct and dominion of an Intelligent and
> Powerful Being ...   This Being governs all
> things, not as a mundane soul, but as the Lord
> of all creatures; who on account of His
> dominion over them is usually stiled the Lord
> God or Supreme Governor of the Universe."

Whiston was anxious that the readers of his Astronomical Principles of Religion (1717) would associate the
name of Newton with the sentiments expressed.   He
regretted that the Scholium Generale was not included in
some of the expositions of Newton's work.

> "In 1728, Dr. Henry Pemberton published his view
> of Sir Isaac Newton's Philosophy.   I would fain
> have had him added those famous scholia or
> Corrolaries of his which are of the greatest
> value for the support of natural and revealed
> religion, but I could not persuade him to it."

So anxious was Whiston to make public these Corollaries to the Principia that in 1729 he published them as a
small tract in 4to and 8vo under the title, Sir Isaac

Newton's Corollaries from his Philosophy in his own Words
(1729).

Some interesting speculations are included in the
Astronomical Principles of Religion on the extent ~~of the~~
~~extent~~ of the Universe.

> "... we have taken a short imperfect view of the
> vastly numerous, the vastly great, and vastly
> distant systems of the fixed stars, or to us
> new systems of worlds quite remote from this our
> planetary or cometary world: In comparison of
> all which systems of worlds, our entire system,
> with its sun, and all its planets and comets,
> must perhaps be inconsiderable; probably not
> the 10,000th, perhaps not the 100,000th, or
> 1,000,000th part of the whole."   (97)

The portrait facing this page shows Whiston holding
a mathematical drawing of the approach of a comet to the
sun.*   This painting was photographed at Lyndon Hall,
Rutland on 29th March 1973 by kind permission of Mr. and
Mrs. John Conant.   It is a hitherto unknown portrait of
Whiston possibly dating from the early decades of the
eighteenth century when he was concerned in writing works
like Astronomical Principles of Religion (1717) which drew
together many of his ideas in the fields of science,
religion and the history of antiquity.

---

*   The artist is uncertain, but a pencil note on the frame
    gives the name Denman.

CHAPTER 4 — REFERENCES

1. Whiston, W. Astronomical Lectures (1715), p.13.

2. Ibid., p. 192.

3. Biographia Britannica (1766), p. 4204.

4. Whiston, W. Memoirs of the Life and Writings of Mr. William Whiston (1749), p. 131.

5. Heath, T.L. Euclid's Elements. Second edition. Dover Publications (1956), p. 2.

6. Osborne, T. A Catalogue of the Libraries of Sir Thomas Burnet (1754), p. 64. Vatican Library Z 999.07.

7. Waterland, D. Advice to a Young Student. Third edition (1760), p. 19.

8. Op. cit. (Reference 5), p. 5-6.

9. Op. cit. (Reference 4), p. 135.

10. Op. cit. (Reference 7), p. 29.

11. Op. cit. (Reference 4), p. 135.

12. Cotes, R. Hydrostatical and Pneumatical Lectures. Second edition (1747), Preface.

13. More, L.T. Isaac Newton. Dover Publications (1962), p. 528.

14. Op. cit. (Reference 4), p. 36.

15. Berry, A. A Short History of Astronomy. Dover Publications (1961), p. 361.

16. Op. cit. (Reference 1), p. 86.

17. B.M. Burney MS. 522.F.2.

18. Hevelius, J. Epistolae IV. De Motu Lunae Libratorio
    Epistolae II. Dr Motu Lunae Libratorio in certas tabulas redacto. Gedani (1654).

19. Newton, I. Philosophiae Naturalis Principia Mathematica (1687).

20. Op. cit. (Reference 1), p. 141.

21. Ibid., p. 169.

22. Ibid., p. 190-191.

23. Ibid., p. 243-246.

24. Ibid., p. 250.

25. Ibid., p. 253.

26. Hevelius, J. Venus in sole visa 1639.   Gedani (1662).

27. Op. cit. (Reference 1), p. 297.

28. Ibid., p. 294.

29. Royal Society Fo.4.28.   Whiston to Folkes, March
      12th 1748.

30. Op. cit. (Reference 1), p. 335.

31. Ibid., p. 344.

32. Cohen, I.B.   Introduction to Reprint Edition of 1728
      edition of Whiston's Astronomical
      Lectures.   Johnson Reprint Corporation
      (1972).

33. Whiston, W.   Sir Isaac Newton's Mathematic Philosophy
      (1716), p. 1.

34. Op. cit. (Reference 7), p. 27.

35. Whiteside, D.T.   The Mathematical Papers of Isaac
      Newton, Volume 3.   C.U.P. (1969),
      (xx-xxvii).

36. Op. cit. (Reference 33), p. 25.

37. Ibid., p. 173.

38. Ibid., p. 195.

39. Op. cit. (Reference 19), Book III, Prop. 35.

40. Op. cit. (Reference 33), p. 238.

41. Ibid., p. 294, 297.

42. Ibid., p. 322-323.

43. Ibid., p. 438-439.

44. Ibid., p. 413.

45. Beinecke Library. Yale University.   Ezra Stiles
      to W. Whiston, May 9th 1751.

46.   Op. cit. (Reference 4), p. 135-136.

47.   Hauksbee, F. Physico-Mechanical Experiments (1709),
          Preface.

48.   Mayor, J.E.B.   Cambridge under Queen Anne (1911).

49.   Ditton, H.   The General Laws of Nature and Motion
          (1705), Preface.

50.   Op. cit. (Reference 4), p. 302.

51.   Nicholson and Rousseau.   This Long Disease, My Life
          Princeton (1968), p. 146.

52.   Ibid., p. 137-138.

53.   B.M. 537 1.23 (4), pp. (3).   London.   Date uncertain.

54.   Ibid.

55.   Whiston and Hauksbee.   A Course of Mechanical,
          Magnetical, Optical, Hydrostatical and
          Pneumatical Experiments (1714).

56.   Dictionary of National Biography.   Vol. XLIII, p.234.

57.   B.M. Catalogue of Printed Maps.   23 (3).

58.   Op. cit. (Reference 4), p. 239.

59.   Op. cit. (Reference 51), p. 143-144.

60.   Whiston, W. The Astronomical Year 1736 (1737), p. 11.

61.   Op. cit. (Reference 51), p. 144.

62.   Worster, Benjamin.   A Compendious and Methodical
          Account of the Principles of Natural
          Philosophy (1722).

63.   Royal Society.   Journal Book Copy. 11. p. 46, 47.

64.   Whiston, W. The Copernicus Explained (1715).

65.   Op. cit. (Reference 4), p. 239.

66.   B.M. 533.e.24. (7) copy of Whiston W.   A New Method
          for Discovering the Longitude both at
          Sea and Land (1715).

67.   Whiston, W. The Calculation of Solar Eclipses without
          Parallaxes (1724), p. 78.

68.   Flamsteed Correspondence.   R.G.O. November 30th
          1714.   Flamsteed to Whiston.

69. Field MS. F.104. Allin to Whiston, 24th April 1715.

70. Whiston, W. An Account of a Surprising Meteor seen
             in the Air 1716 (1716).   B.M. copy.
             Shelf No. 873.1.2.(1).

71. Ibid., p. 57-59.

72. Op. cit. (Reference 70) second edition.
             MS. notes on p. 67.   B.M. copy.
             Shelf No. 873.1.2.(1).

73. Philosophical Transactions, No. 135.

74. Ibid., No. 341.

75. Whiston, W. An Account of a Surprising Meteor seen
             in the Air, March 19th 1719. (1719), p. 24.

76. Ibid., p. 27.

77. Ibid., p. 43.

78. Royal Society. Record Book Copy. Vol. XI,
             p. 156-159.

79. Royal Society. Journal Book Copy. XII. October 26th
             1721, p. 155.

80. Royal Society. Register Book Copy. Vol. XI, p. 158.

81. Op. cit. (Reference 67), p. 72, 73.

82. Royal Society. Journal Book Copy. Vol. XII,
             p. 515,516.

83. Ibid., Vol. XIII, p. 500-501.

84. Op. cit. (Reference 60), p. 21.

85. Ibid., p. 24.

86. Ibid., p. 25-26.

87. Gentleman's Magazine, Volume XII, p. 106.

88. Op. cit. (Reference 29).

89. Op. cit. (Reference 45).

90. B.M. Map Room. 148.e.3.(1).   Whiston, W.   A Scheme
             of the Solar System.

91. B.M. Map Room. 23 (3).   Whiston, W.   A Calculation
             of the Great Eclipse of the Sun, April
             22nd, 1715.

92. B.M. Map Room. 26 (2) Whiston, W. The Transits of Venus and Mercury over the Sun at their Ascending and Descending Nodes for two Centuries and a Half (1736)

93. B.M. Map Room. 1030.(37). Whiston, W. A New Map of Europe (1733).

94. Whiston, W. Astronomical Principles of Religion (1717), p. 92.

95. Ibid., p. 95.

96. Ibid., p. 141.

97. Ibid., p. 253.

APPENDIX

SCIENTIFIC PUBLICATIONS OF WILLIAM WHISTON

1696

A New Theory of the Earth from its original to the Consumma-
tion of all things wherein the Creation of the World in Six
Days, the Universal Deluge and the General Conflagration, as
laid down in the Holy Scriptures are shewn to be perfectly
agreeable to Reason and Philosophy.
 Second edition 1708, Third edition 1722,
 Fourth edition 1725, Fifth edition 1737
 Sixth edition 1755.

1698

A Vindication of the New Theory of the Earth from the
exceptions of Mr. Keill and others.

1700

A Second Defence of the New Theory of the Earth from the
exceptions of Mr. John Keill.

1703

Andrea Tacquet Elementa Euclidea Geometriae Planae ac Solidae
et Selecta ex Archimede Theoremata. (Ed.)
 Latin editions 1710, 1722, 1795.
 Latin editions (Venice)1737, 1781.

1707

Arithmetica Universalis. (Ed.)

Praelectiones Astronomicae Cantabrigiae in Scholis Publicis
Habitae quibus Accedunt Tabulae Plurimae Astronomicae
Flamstedianae Correctae, Halleianae, Cassinianae et
Streetianae.

1710

Praelectiones Physico-Mathematicae Cantabrigiae in Scholis
Publicis Habitae.
 Second Latin edition 1726.

1712

A Scheme of the Solar System.   (Chart)

1714

The Elements of Euclid with select Theorems out of Archimedes
by the learned Andrew Tacquet.   (Ed.)

To which are added:

Practical corollaries, shewing the uses of many of the
propositions.
        English translation of 1703 Latin edition.
        English editions 1714, 1719, 1727, 1753.
        English edition (Dublin) 1728.

Reasons for a Bill Proposing a Reward for the Discovery of
the Longitude.   (Broadsheet) (with Humphrey Ditton).

A New Method for Discovering the Longitude both at Sea and
Land.   (with H. Ditton)
        Second edition 1715.

The Cause of the Deluge Demonstrated.

A Course of Mechanical, Magnetical, Optical, Hydrostatical
and Pneumatical Experiments.

1715

Astronomical Lectures Read in the Public Schools at Cambridge
        Whereunto is added a collection of Astronomical
        Tables being those of Mr. Flamsteed corrected,
        Dr. Halley, Monsieur Cassini and Mr. Street.
        This work is the English translation of
        Praelectiones Astronomicae (1707).
        Second English edition 1728.

A Calculation of the Great Eclipse of Sun, April 22nd 1715.
        (Chart)

An Account of the Eclipses of the Sun 1715 and 1724   (Chart)

The Copernicus Explained.

1716

Sir Isaac Newton's Philosophy more Easily Demonstrated.
        with Dr. Halley's Account of Comets illustrated
        being forty lectures read in the public schools
        at Cambridge.   This is the English translation
        of Praelectiones Physico-Mathematicae Cantabrigiae
        (1710).

1716 (Continued)

An Account of a Surprising Meteor seen in the Air, March 6th 1716.

1717

Astronomical Principles of Religion.
      Second edition 1725.

1719

The Longitude and Latitude found by the Inclinatory or Dipping Needle.
      Second edition 1721.
      This included a Historical Preface.

An Account of a Surprising Meteor seen in the Air, March 19th, 1719.

1723

A Scheme of the Transits of Mercury and Venus over the sun for two and a half centuries.
      Second edition 1736.

1724

A Scheme of the Solar Eclipse, May 11th 1724.

The Calculation of solar eclipses without parallaxes.
      With a specimen of the same in the total eclipse
      of 1724.   To which is added a proposal how, with
      the latitude given, the geographical longitude of
      all the parts of the earth may be settled by the
      bare knowledge of the duration of solar eclipses.
      With an account of some late observations made
      with the dipping needles.

1729

Sir Isaac Newton's Corollaries from his Philosophy in his own words.

1733

A Map of Europe: with the course of eight eclipses of the Sun drawn upon it.   (Chart)

1737

The Memorial of William Whiston, Clerk, sometime Professor
of Mathematics in the University of Cambridge. (Broadsheet)

The Astronomical Year or an Account of the many remarkable
celestial phenomena of the great year 1736, particularly
of the late comet which was foretold by Sir Isaac Newton
and appeared at its Conclusion.

The Astronomical Year 1736.

1738

The Longitude Discovered by the Eclipses, Occultations and
Conjunctions of Jupiter's Planets.
        Some copies of this work are preceded by an
        Historical Preface dated Aug. 29th 1741.

1743

An Exact Trigonometrical Survey of the British Channell
from the North Foreland to the Scilly Islands and Cape Clear
on the southwest part of Ireland.
        Chart 815 x 1670 mm.

C H A P T E R   F I V E

WHISTON'S RELIGIOUS WORKS

5.1   Background to his Religious Beliefs.

5.2   Studies in Biblical Chronology.

5.3   Interpretation of Scripture Prophecies.

5.4   History of the Doctrine of the Trinity.

5.5   Whiston's Interest in the Doctrines of the
       Early Church.   His Expulsion from the
       University of Cambridge.

5.6   The Nature of Whiston's Heretical Views.

5.7   The Society for Promoting Primitive Christianity
       and the Practice of Infant Baptism.

5.8   Whiston's Primitive Catechism and His Work
       for the Charity Schools.

5.9   Correspondents of Whiston regarding Primitive
       Christianity.

5.10   Views of Modern Biblical Scholarship regarding
       the Apostolical Constitutions.

Appendix

       Chronological Table of the significant
       religious publications of William Whiston.

CHAPTER FIVE

WHISTON'S RELIGIOUS WORKS

Whiston was deeply involved in the religious movements of his time. His writings on religion fall into three classes:

(a) Studies in biblical chronology.

(b) The interpretation of scripture prophecies.

(c) The study of the writings of the primitive church. This in turn led to his involvement in the revival of a form of Arianism.

5.1 BACKGROUND TO HIS RELIGIOUS BELIEFS

As already pointed out in Chapter One, Whiston seems to have absorbed from his father and grandfather a Puritanical outlook in religious matters. This manifested itself in a strict code of moral behaviour which was condemnatory of pleasure and enjoyment of the good things of life. His life was lived under the stern and penetrating gaze of a severe Deity. Long prayers and sermons marked the passing of time and these were full of warnings and forebodings about the punishments awaiting those whose life could be considered dissolute or vain.

The radical attitude which Whiston adopted in his attitude to the practice of Christianity, and his desire to go back to primitive sources, probably has its origin in the religious movements of the Reformation.

Erasmus advocated the idea of the restoration of primitive Christianity. Zwingli (Ulrich) enthusiastically took up the same attitude as Erasmus in seeing the externalisation of religion as a great abuse. A group of Zwingli's circle became known as the Anabaptists. They aimed at the restoration of Primitive Christianity and the adoption of the Sermon on the Mount as a literal code for all Christians. Baptism of infants was replaced by baptism in mature years. Anabaptists in Zurich were subjected to the death penalty in 1525. Many persons fled from the centre of the dispute with a consequent dissemination of the ideas to other parts of Europe.

The Puritans of the early seventeenth century had high ideals of integrity and of service to the community. Their preachers taught the equality of all men, a doctrine carried to its extreme by the Levellers during the Civil War and Commonwealth. The renunciation of privilege and the insistence of social levelling had been advocated by members of Luther's immediate circle, particularly by Carlstadt. One good man was as good as another and better than a bad peer or bishop or king. If men honestly searched their conscience they could not fail to agree about God's will.

Three aspects of Puritanism can be traced in Whiston's attitudes namely the emphasis on preaching as against sacramental and liturgical functions; the enforcement of a strict code of discipline as a means to combat idleness and to encourage devotion to one's duty; Sabbatarianism and its association with preaching, Bible reading and household prayers.

In the early part of the century, the demand for
sermons and religious discussion was almost insatiable.
It was a sign of the awareness of men of a spiritual crisis
in their society.   Archbishop Laud's emphasis on ritual
and ceremony seemed like a return to the practice of Roman
Catholicism before the Reformation.   In 1641 the House of
Commons made it lawful for a parish "to set up a lecture and
to maintain an orthodox minister at their own charge to
preach every Lord's Day ...."

In spite of many proposals to adopt measures whereby
the clergy would enjoy a more equal share of the Church's
wealth, the hierarchy continued to authorise pluralism.
The abuse of pluralism resulted in a minority of clergy and
most Bishops enjoying handsome revenues while leaving some
congregations totally unprovided for.   In 1603, according
to the Bishops' figures, nearly sixty per cent of the
benefices were occupied by persons either too stupid or
too politically unreliable to be allowed to preach.   The
worst pluralists were the Bishops and the clergy of the
cathedrals, universities and court.

The most obviously political religious group of the
Interregnum was that of the Fifth Monarchists, so called
because of their belief that the time of the fifth monarchy
was at hand which would succeed the Assyrian, Persian,
Greek and Roman eras.   They believed that the reign of
Christ on earth with the saints for a thousand years was
about to begin.   The sense of the imminence of a new
spiritual epoch in which God's people should be free in a

new way, was one of the many Fifth Monarchist concepts
which the Quakers took over.   Whiston's belief in the near
approach of the end of the world places him within this
tradition.   His expectation of the Millenium was based on
the interpretation of Scripture prophecies.   Millenium
sects or movements always picture salvation as:

(a) collective, in the sense that it is to be
enjoyed by the faithful as a collectivity;

(b) terrestrial; it is to be realised on this
earth and not in some other-worldly heaven;

(c) imminent, it is to come both soon and suddenly;

(d) total, it is utterly to transform life on earth,
so that the new dispensation will be no mere improvement on
the present but perfection itself;

(e) miraculous, to be accomplished by, or with the
help of, supernatural agencies.

A letter written in 1748 to the Duchess of Portland
reads:

> "Mr. Whiston was with me this morning and assured
> me that eighteen years hence the Jews will be
> converted, and that twenty years hence the
> Millenium will begin."   (1)

The essence of Puritanism and revolutionary creeds
lay in the belief that God intended the betterment of man's
life on earth, that men could understand God's purposes
and co-operate with him to bring them to fruition and so
men's innermost wishes could be believed to be God's will.
After 1660, the quietist, pacifist tendency increased as
Puritanism turned into nonconformity.

Most Puritan ministers had adopted the traditional

view that God's elect were a minority but in order to encourage their congregations and save them from despair, they taught that anyone who seriously worried about his salvation probably already had sparks of divine grace at work in him.   It was a short but momentous step to proclaim that all men were equally eligible to receive divine grace. Far-reaching consequences followed from this premise.   In 1650 compulsory attendance at one's parish church was legally abolished.   It marked a new type of freedom, the liberation of the common man from servitude to squire and parson.   This tolerance did not extend to Roman Catholics who were regarded as agents of a foreign power.   Whiston was typical of his times in his strong anti-papist attitude.

The influence of the radical Protestant tradition spread far beyond Puritan circles.   An appeal to Scripture or to conscience could be used to call any authority in question.   It seemed possible to find in the Bible a text to prove whatever a person wanted to prove.   After censorship was lifted in the sixteen-forties, lower-class Englishmen discovered in the Bible and in their consciences a wide interpretation of traditional values.

In 1660 Bishops came back to their sees, recovered their lands, and returned to their seats in the House of Lords, but the Church never regained its old position and the Bishops never again had the same dominance in politics. The mediaeval concept of the clergy as a separate estate vanished and they became one of many professions.   Church courts slowly lost their power.   When Whiston's condemnation for heresy is discussed, it will be seen that although

a body of Churchmen were willing to condemn him they lacked
any authority by which to make their condemnation effective.
The existence of dissenting sects meant that the sentence
of excommunication lost much of its efficacy.   The association of extremist behaviour in religious matters with
radical politics led to a reaction which contributed to the
growth of rational religion and deism.   At the time of
Whiston's undergraduate studies in Cambridge 1685-1688,
the prevailing climate was fostering the growth of deism and
religious belief founded on a rational basis.

After William of Orange became King in 1688, the possibility of comprehending Presbyterians within the Church of
England was again mooted.   The Church was rent by divisions.
The Non-Jurors were clergy who, having sworn an Oath of
Allegiance to James, felt they could not conscientiously
accept William and Mary as sovereigns.   Whiston refers to
his plight in searching out a bishop to whom he might apply
for holy orders.

> "I had no mind to apply myself to a Bishop, how
> excellent soever, who had come into the place
> of any who were not satisfied with the Oaths to
> King William and Queen Mary, and so had been
> deprived for preferring conscience to preferment."[2]

Many clergy admired, even if they did not imitate, the
courage and consistency of the non-jurors.

Bishops who resigned in 1689 or died later were
replaced by men who accepted the Revolution Settlement,
including Latitudinarians like John Tillotson who succeeded
William Sancroft as Archbishop of Canterbury.   The term
Latitudinarian, was coined originally as a term of abuse to

designate the Cambridge Platonists. It seemed to Englishmen
of the time to afford a relief from both Puritanism and the
High Churchmanship of Laud. Most were Cambridge men who
had been taught by John Smith, Ralph Cudworth or John Moore.
It was the avowed purpose of the Latitudinarians to
eliminate the irrational from religion; the use of one's
mental powers could only advance the cause of faith. Most
were careful to emphasise that natural religion must be
supplemented by revelation, but they rejoiced in the belief
that the mind, without any appeal to extraneous authority,
could grasp in outline a religion which included all the
essentials of belief.[3]

Religious works by persons of these views are well
represented in Whiston's library:

Stillingfleet	Unreasonableness of Separation. (1681)
Tillotson	Rule of Faith. (1666)
Simon Patrick	Christian Sacrifice. (1682)
Hammond	Reasonableness of Christian Religion. (1650).

John Toland's book, Christianity not Mysterious, (1696),
was a plea for rational religion. John Locke's
Reasonableness of Christianity (1695) was followed by a
series of books attempting to reconcile reason and faith.

The Latitudinarians thought that the authority of
antiquity had been much overrated. They wanted a re-
assessment of the value of Christian antiquity as distinct
from classical authors of the same period. There was a
resurgence of interest in primitive Christian literature
and Whiston was caught up in the movement to widen the

canon of Scripture to include such works as the <u>Apostolical</u>
<u>Constitutions</u>. His translation into English from the
Greek of the works of Flavius Josephus was a contribution
to the historical context of early Christianity.

John Locke had stressed the importance of reason as,

> "natural revelation, whereby the Father of Light
> and fountain of all knowledge, communicates to
> mankind that portion of truth which he had laid
> within the reach of their natural faculties." (4)

For John Toland, the essence of New Testament revelation was that it did really reveal things which were formerly mysterious. Stripping the truth bare of all cere-
monial and symbolical trappings was the preoccupation of
the Deists. The accretions of mythology, paganism and
popery had to go and Protestantism evolved in England into
a form of rationalism which was both anti-pagan and anti-
Catholic. (5)

The Deist controversy falls into two stages. In the
first, the debate concerns nature, reason and the degree
to which Christianity offers anything not already latent
in nature and reason. In its second stage, the issue at
stake was the historical proof of the genuineness of the
Christian records. It seems that Whiston was involved in
both stages of the Deist controversy and the renewal of
interest in primitive Christian literature already mentioned
was one of the outcomes of the debate.

Matthew Tindal's work, <u>Christianity as old as the</u>
<u>Creation</u> (1730) taught that God's work is perfect and
perfectly reveals him. Tindal assumed that from the first,
man was perfectly equipped to grasp this perfect religion.

This was one of the fundamental weaknesses of the Deists.
They lacked a sense of history and oversimplified the
problems of human development.

The Deist controversy and its rise to a climax in
about 1730 corresponded closely to the span of Whiston's
academic career.   The three great opponents of Deism, and
their works which marked its wane were William Law,
The Case of Reason (1731), George Berkeley, Alciphron (1732),
and Bishop Joseph Butler, The Analogy of Religion (1736).
These three scholarly men broke through the restricted frame
in which eighteenth century rationalism moved and pointed
the way to a new understanding of the role of reason.

5.2   STUDIES IN BIBLICAL CHRONOLOGY

Whiston's first publication of a strictly religious
character was his Short View of the Chronology of the Old
Testament and the Harmony of the Four Evangelists (1702).

In his memoirs, Whiston tells us:

"I entirely followed at first the Masorete Hebrew
Copy and its numbers, which I then took to be
most authentic; but because on further enquiry,
I afterwards altered my mind as to that matter,
and fully satisfied myself the Samaritan
Pentateuch as well as Josephus's Copy of the
Hebrew, together with the Septuagint Version,
and the most authentic records of heathen
antiquity, agree in a chronology that lengthens
the interval since the deluge about 580 years." (6)

Whiston's chronological studies were basically a
revision of the work of Bishop James Ussher (1581-1656) who
during the years 1650-1654 published his Annales Veteri et
Novi Testamenti.   The dates from this work were inserted in
the margin of the Authorised Version and were current for

two centuries.

Correspondents of Whiston in his chronological studies included Daniel Whitby, Fellow of Trinity College, Oxford, and William Lloyd, Bishop of Lichfield.[7]   Whiston's chronological studies were known to Peter Allix (1641-1717), pastor of the Reformed Church of France and biblical scholar. Whiston was involved in controversies with Allix on chronological matters and also with the revival of Arianism.

In 1721, Whiston published A Chronological Table of the Hebrew, Phoenician and Chaldean Antiquities.*   In 1727 and 1728, a two volume work, A Collection of Authentic Records belonging to the Old and New Testament was published; The Horeb Covenant Revived (1730) and A Description of the Tabernacle of Moses and of the Temples at Jerusalem (1731).   The latter work was a large sheet of illustrations.

Whiston lectured on the Tabernacle of Moses and the Temple of Jerusalem during 1728 and 1729.

> "In the year 1726, I procured to be made me by
> Mr. Crosedale, a very skilful workman, but
> according to my own directions and at the
> expense of about forty guineas, a model of the
> Tabernacle of Moses and of the Temple at
> Jerusalem ... and had lectures upon that at
> London, Bath and Tunbridge Wells."   (8)

Whiston presented himself at Court to show his model of the temple to Queen Caroline,

> "... he (Whiston) had lately been at Court to show
> her a model of the New Jerusalem as described
> by the prophet Ezechiel."   28th July 1730.   (9)

---

*   This table was meant as an appendix to the work An Essay
    towards Restoring the True Text of the Old Testament
    (1722).   I have not been able to trace a copy of this
    table of Antiquities.

In 1730 William and George Whiston, sons of William Whiston, published proposals for printing by subscription Mosis Chorenensis Historiae Armeniacae Libri III. This work appeared in 1736. It was without doubt due to their father's interest in the chronology of antiquity that they undertook this task.

Mosis Chorenensis wrote a history of Armenia which appeared about A.D. 402. The first book dealt with the state of Armenia from the dispersion of Babel to Alexander the Great; the second from Alexander the Great to the death of their King Tiridates about A.D. 300; the third from A.D. 300 to the middle of the fifth century. It was printed by an Armenian Archbishop in Amsterdam in 1695. William and George Whiston were the first to translate it into any other language. Their publication gave the Armenian and Latin text of the work. The expense of the Armenian type (the first in England) was met by the donations of several gentlemen.

Whiston used Mosis Chorenensis as an authority for events in the reign of Sennacherib and other happenings of Jewish antiquity. This is indicated by Whiston's MS. notes in H. Prideaux The Old and New Testament connected in the History of the Jews and Neighbouring Nations (1718).[10]

Mosis Chorenensis Historiae Armeniacae was published by John Whiston and included two appendices: Mosis Chorenensis Geographia and Epistolae duae Armeniacae. The work is preceded by a list of subscribers among whom appear many friends and patrons of William Whiston. This work is attractively presented and includes an explanation of

Armenian letters and numerals and a map of Armenia.

The two Armenian epistles included in this work were what Whiston believed to be An Epistle of the Corinthians to St. Paul, and St. Paul's answer.   In Whiston's A Collection of Authentic Records belonging to the Old and New Testament (1728), he devotes about 50 pages to the text of these epistles and gives reasons for their accept-ance as genuine.[11]

Whiston's major contribution to biblical chronology appeared in 1737, The Genuine Works of Flavius Josephus, translated from the original Greek of Havercamp's accurate edition.[12]   It consists of a translation of the twenty books of Jewish Antiquities, the life of Josephus, seven books of the Jewish War and two books against Apion.   It is illustrated with plans and descriptions of Solomon's, Zorobabel's, Herod's and Ezekiel's Temples with maps of Judea and Jerusalem.   The edition is annotated, parallel texts of Scripture are given and Whiston's estimate of the chronology of events is given in the margin.

This translation of Josephus occupied more than two years of Whiston's efforts.   He kept some notes about it at the back of the Minute Book of the Society for Promoting Primitive Christianity,

> "I began this version on December 9th AD 1734,
> the day I was 67 years of age and finished it on
> January 6th 1737 in the beginning of my 70th
> year having been about two years and one month
> about it."   (13)

The importance of Flavius Josephus is that he was a Jewish historian in the first century of Christianity.

He was involved in negotiations with the Roman authorities during the reign of Nero. He eventually became a Roman citizen in the reign of Vespasian and received an estate in Judea. Josephus mentions John the Baptist and Christ and introduces scriptural authorities as well as facts gleaned from Greek histories.[14]

Whiston saw Josephus as an authority from whose writings, the chronology of Jewish history and early Christianity could be calculated. Two long letters from October and November 1742 survive between John Jackson, the author of Chronological Antiquities, and Whiston, which deal largely with chronological details mentioned in Josephus and characters in the story of early Christianity. This is an example of the kind of problem examined:

> "You remember that lately at my house you told
> me that Ananias, called the High Priest (Acts,
> 23 and 24) was not really the High Priest but
> Jonathan as you thought appeared from Josephus.
> I have since examined the matter and find it very
> plain that Josephus agrees with St. Luke that
> Ananias and not Jonathan was then the High Priest
> and wonder how you came to suppose otherwise
> having read Josephus so carefully."   (15)

Whiston's Josephus went through numerous reprintings in the eighteenth and nineteenth centuries.*

In 1745, Whiston published a three volume work, Mr. Whiston's Sacred History of the Old and New Testament. This is a revision of Humphrey Prideaux's work, The Old and New Testament connected in the History of the Jews and the Neighbouring Nations (2 vols.) (1716-1718). As a result of

---

* Whiston's Josephus was the principal version available in English until it was superficially revised by A.R. Shilleto in 1889.

his work on Josephus, Whiston felt able to make considerable
corrections to Prideaux's work.   A manuscript note
indicates that Whiston considered using as an authority on
chronology the Astronomical Canon of Ptolemy,

> "beginning the first New Years Day Feb. 26 Anno
> 747 before the Christian era with both the reign
> of Nabonassar, King of Babylon.   This canon is
> confirmed along by eclipses of the sun and moon,
> actually observed at Babylon by a College of
> Astronomers who came long ago thither from Egypt
> and was continued afterwards by the astronomers
> of Alexandria ...   The year it uses is 365 days
> only, without any quarter of a day...   It was
> first discovered in the reigns of Queen Elizabeth
> and King James I and having in the original the
> sums of years at every reign as well as the
> years of the reigns themselves.   They are lyable
> to no mistakes in number which is an inestimable
> advantage of Chronology."*

5.3   INTERPRETATION   OF   SCRIPTURE   PROPHESIES

In 1707 Whiston was invited to deliver the series of
sermons known as the Boyle lectures.   This lectureship
dated from 1692 and was endowed by the Honourable Robert
Boyle who had died in 1691 and bequeathed £50 per annum as
the salary of some London divine whose chief duty would be
to preach eight sermons each year in various London churches,

> "for proving the Christian religion against notor-
> ious infidels, viz. Atheists, Theists, Pagans,
> Jews and Mahometans, not descending lower to any
> controversies, that are among Christians
> themselves."   (17)

Richard Bentley (1662-1742) the eminent English

---

* The British Museum has Whiston's copy of Prideaux's
  The Old and the New Testament connected in the History
  of the Jews and Neighbouring Nations (1718).   This is
  evidently the copy which he used to prepare his revision
  of the work in 1745.   It is covered with manuscript
  notes, additions, corrections and directions for omission
  of passages. (16)

classicist and Master of Trinity College, Cambridge, gave
the first series of Boyle lectures.   The last three
lectures of this series bore the title: A Confutation of
Atheism from the Origin and Frame of the World and were
a skilful elaboration of Newton's physical system of the
world, using these ideas as a support for the "argument from
design" for belief in a Providential Deity and in His wisdom
and goodness.

Samuel Clarke (1675-1729) had been a close associate
of Whiston since 1697.   In the years 1704 and 1705, he
delivered the Boyle lectures under the titles of
A Demonstration of the Being and Attributes of God and
A Discourse concerning Natural and Revealed Religion.   It
was about this time that Clarke and Whiston began to share
their common interest in the primitive writers of
Christianity and both men began to suspect that the
Athanasian doctrine of the Trinity was not the doctrine of
the early church.   Clarke was considered the foremost
metaphysician in England; he is best remembered as the
correspondent, on behalf of Queen Caroline, with Leibnitz
on the subject of Newtonian philosophy.

In 1707, there was considerable interest in the interpretation of Scripture prophecies because certain French
prophets were claiming current events as the fulfilment
of these prophecies.   Whiston's Boyle lectures were
entitled The Accomplishment of Scripture Prophecies.   He
prefaced his sermons by a series of observations which he
believes will help in the understanding of the ancient
prophecies.

His first observation deals with the interpretation
to be given to the terms day and year in the scriptures and
agrees closely with that given in his New Theory of the
Earth (1696) (see Chapter Two).   He also points out that
a rational interpretation has been found for many scripture
prophecies which were thought at one time to be enigmatic.
One can hope that eventually all scripture prophecies will
be clearly interpreted.   His final observation is that
each scripture prophecy is capable of only one interpretation.
He maintains that a double sense of the interpretation of
scripture is not to be found in the apostolic tradition.
Whiston's Boyle sermons consist of an exposition of 23
prophecies covering such major events in the history of
salvation as the foretelling of the birth of the Messias,
the waters of the deluge, the destruction of the world by
the waters of the deluge, the children of Israel as des-
cendants of Abraham, the fall of Jericho.   Whiston studies
all these prophecies and their fulfilment in an almost
chronological order and shows great skill in using his
chronological studies to show the exact time of the fulfilment of these prophecies by particular events recorded in
biblical history.   To cite an example, the fall of Jericho
(Prophecy III) was foretold in the year 1491 before Christ
and came to fulfilment in the year 918 B.C.[18]

Whiston warns against the French prophets who,
as already mentioned, were active in these years.

> "If any person in this age, who pretend to a
> prophetic spirit do foretell events, whether of
> mercy or of judgment, which do not come to pass
> according, we have the warrant of God himself for
> their rejection."   (19)

Whiston's attitude to prophecy was within the
Latitudinarian position of defending revelation against its
detractors.　　Miracles and prophecy were considered to be
the two main pillars on which revelation rested.　　The purpose of prophecy was considered to be an intimation from
God of things yet to come.　　An immense amount of learning
and ingenuity was expended in examining the evidence hidden
in Scripture prophecies.　　Other writings in the same vein
include:

J. Jackson	Address to the Deists (1744)
E. Chandler	A Defence of Christianity from the prophecies of the Old Testament (1725)
M. Lowman	The Argument from Prophecy Vindicated (1733)

The Deists exalted reason and nature and by implica-
tion challenged the validity of revelation.　　Anthony
Collins was the author of two works which reiterated reason's
right to pronounce on the contents of revelation.　　These
were, A Discourse of Free Thinking (1713) and A Discourse of
Grounds and Reasons of the Christian Religion (1724).　　This
latter work questioned whether miracles and prophecy could
be trusted as the foundations of revelation.

Collins' strategy was first to prove that the fulfilment of prophecy is the chief guarantee of the truth of
the Christian religion and then to expose the weakness of
the appeal to prophecy.　　He undertook to show that none of
the Old Testament prophecies to which apologists so con-
fidently appealed had been fulfilled in an exact or literal
sense.

Whiston and Collins represent controversialists on either side of the Deist fence. The Deists were trying to evolve a critical approach to the Bible but they lacked the necessary tools of historical understanding and textual skill. On the other hand, Whiston's historical understanding was limited by the traditional chronological studies which demanded a literal and unique exegesis of biblical texts.

In 1724-25, Whiston wrote an answer to Collins entitled The Literal Accomplishment of Scripture Prophecies. The plate opposite is the title page of this work. See Appendix to this chapter for a discussion of its contents.

5.4 HISTORY OF THE DOCTRINE OF THE TRINITY

It is not true to say that the early church fathers had no clear conception of the Trinity. The anti-gnostic fathers, Irenaeus, Hippolytus and Tertullian did believe in a triune God. Their formulation of belief in the doctrine of the Trinity was not phrased in the same way as that with which we have become familiar. Although Tertullian was the first to use the term Trinity, he did not use this term in the commonly accepted sense. Among the early Church Fathers, there is little mention of the Spirit; the work of the Spirit was widely assumed to be carried on by the Logos.

The Alexandrian fathers, Clement of Alexandria and Origen both represent the theology of the East which is much more speculative than that of the West. Clement did not try to explain the relation of the Holy Spirit to the Father

The LITERAL

ACCOMPLISHMENT

OF

Scripture Prophecies.

Being a full Anfwer to a late Difcourfe, *Of the Grounds and Reafons of the Chriftian Religion.*

CONTAINING

I. Predictions of the Old Teftament concerning the Meffias, cited in the Four Gofpels and *Acts of the Apoftles:* with their literal Completions in Jefus of *Nazareth.*

II. Predictions in *Dan* xi. from the Days of *Cyrus* the *Perfian,* to the End of the World: With their Literal Completions all along, to this Day.

III. Predictions contained in the Four Gofpels: With their Literal Completions all along, to this Day.

IV. Natural Preparations for the Deftruction of Antichrift; for the Revival of Primitive Chriftianity; and for the Reftoration of the *Jews,* in the laft Days.

V. Predictions preparatory to the Deftruction of Antichrift; to the Revival of Primitive Chriftianity; and to the Reftoration of the *Jews,* in the laft Days. With the Literal Completions of thofe whofe Periods are paft; and Conjectures concerning thofe that feem to relate to the prefent Age.

VI. Corrections and Improvements to the Sermons at Mr. *Boyle's* Lecture, concerning the *Accomplifhment of Scripture Prophecies.*

VII. Corrections and Improvements to the *Effay on the Revelation of St.* John.

With a Large *Appendix,* to prove, that *Arifteas's* Hiftory of the Verfion of the *Pentateuch,* by the LXXII. Interpreters, ftill extant, is genuine.

To which are added, *Propofals* for printing *Authentick Records* concerning the *Jewifh* and Chriftian Religions, *&c.*

By *WILL. WHISTON,* M. A. Sometime Profeffor of the Mathematicks in the Univerfity of *Cambridge.*

LONDON: Printed for J. SENEX, in *Fleetftreet;* and W. TAYLOR, in *Pater-Nofter-Row.* MDCCXXIV

and the Son.   Origen occupied a teaching role as a catechist
in Alexandria after Clement's time as Bishop.   Origen had
absorbed many neo-platonic ideas into his teaching and
maintained that one God the Father reveals Himself and His
works through the Son (the Logos).   The Father is so much
apart from created things that only through His Son can He
reveal Himself to the world.

The controversy between Arius and Athanasius which
came to a head at the Council of Nicea 325 A.D. was not in
fact concerned with the doctrine of the Trinity as we understand it today.   The debate centred around the Divinity
of the Son.   Arius' idea of God was monarchic.   The Son
of God did not exist alongside God from the beginning.
There was a time when He was not in existence.   The Logos
had a beginning in that He was generated by the Father.
Arius was opposed by his own Bishop, Alexander, and also by
Athanasius.   Athanasius strongly emphasised the unity of
God while still recognising three distinct persons
(hypostases).   It was Athanasius who favoured the use of
the term HOMOOUSIOS to signify that the Son was of the same
substance as the Father.

At the Council of Nicea, there were not just two fields
of thought characterised by Arius and Athanasius.   There
were many others, in fact a majority, including many
prominent persons, one of whom was Eusebius of Caesarea.
These persons opposed the use of the term HOMOOUSIOS.   They
would have preferred the use of the term HOMOIOUSIOS which
allowed that the Son was of similar substance to the Father,
but did not agree that the Son was of the same substance as

the Father.   They did not consider the term HOMOOUSIOS to
have a foundation in biblical tradition.

Eusebius was in trouble before the beginning of the
Council of Nicea.   He had already been condemned for his
ideas at the small council of Antioch.

The Council of Nicea adopted the following statement:

> "We believe in one God, the Father Almighty,
> maker of all things visible and invisible.   And
> in one Lord Jesus Christ, begotten, not made,
> being of one substance with the Father."

Athanasius wrote a full treatise in defence of the use of
the term HOMOOUSIOS.   Eusebius wrote a letter to his Church
at Caesarea explaining the sense in which the term HOMOOUSIOS
is to be understood.   This letter is not usually printed
with the works of Eusebius and so there can be misunder-
standing of the position which he finally adopted.

Basil the Great, Gregory of Nyssa and Gregory of
Nazianzus were converts to the use of the term HOMOOUSIOS.
By their emphasis on the three hypostases in the divine
Being, they freed the Nicene doctrine from any taints of
Sabellian ideas.   Sabellianism was a modalistic monarchic
doctrine of the Trinity which saw the Trinity as different
manifestations of the one Godhead.

The dispute about the Divinity of the Holy Spirit did
not arise until 359 A.D.   A community of Egyptian Christians
accepted the divinity of the Son but rejected the Divinity
of the Spirit.   Their problem was presented to Athanasius
by one of their leaders, Serapion.   In his Epistola ad
Serapion, Athanasius uses the same arguments about the Spirit
that he used against the Arians when speaking about the Son.

He reminded the Christians that they were baptised in the name of the Father and of the Son and of the Spirit. The Spirit sanctifies and therefore cannot be a thing; the Spirit must be God.

Those who resisted the doctrine of the divinity of the Spirit were known as PNUMATOMACHIANS (Spirit fighters). They had two astute theologians as spokesmen, Aetius and Eunomius. The Cappadocians finally clarified the issue of the divinity of the Spirit by explaining the hypostasis of persons.

The Council of Constantinople in 381 A.D. approved the formula,

> "And we believe in the Holy Spirit, the Lord, the
> life-giving, who proceeds from the Father and who
> is to be glorified with the Father and the Son
> and who speaks through the prophets." (20)

The Western conception of the Trinity reached its final statement in Augustine's great work De Trinitate.

Augustine compared the two processions of the Divine Life (later known as Filiation and Spiration) to the analogical processes of human self-knowledge and self-love.* His conception of the generation of the Son as the act of thinking on the part of the Father was based on Tertullian. The explanation of the Spirit as the mutual love of the Father and the Son was the fruit of Augustine's own reflections. This so-called psychological theory of the Trinity was taken over by medieval scholastics and developed by them. The doctrine of the Trinity received classical

---

* The term 'procession' is here used in the theological sense of the coming forth from a source.

exposition in the writings of St. Thomas Aquinas.

Augustine explained the 'processions' of the persons of the Trinity.

In 580 The Council of Toledo endorsed the use of the term FILIOQUE in the Constantinopolitan formula and thus affirmed the divinity of the three persons.

5.5 WHISTON'S INTEREST IN DOCTRINES OF THE EARLY CHURCH. HIS EXPULSION FROM THE UNIVERSITY OF CAMBRIDGE

Whiston tells us that in February 1708, he was asked by a friend to outline a method for the study of divinity. The method proposed involved a close study of the doctrine and faith of the first Christians concerning the Trinity. At this stage Whiston acknowledged that the original teaching on the Trinity was very different from that commonly held, and he decided to re-examine the whole matter from the New Testament and the most ancient genuine extant sources.[21]

Whiston's first confession of doubt in the proper eternity and omniscience of Christ dates from 1706. In August 1708, Whiston presented to Dr. Lany, Vice-Chancellor at Cambridge his Essay upon the Apostolic Constitutions. Dr. Lany refused the licence for the printing of this work. In 1709 Whiston printed in London his work Sermons and Essays Upon Several Subjects to which was added Novatian's work De Trinitate. It was this publication together with the doxology at the end of his sermon before the governors of the Charity schools and his proposals for printing Primitive Christianity Revived which led to his dismissal from

Cambridge in 1710.

On Sunday, 22nd October 1710, Whiston was summoned and conducted into the lodge of the Vice Chancellor, Dr. Roderick, Provost of King's College. Heads of the other colleges, making a majority of the whole, were gathered. He was shown his book <u>Essays and Sermons on Several Subjects</u> (1709), also the doxology at the end of his sermon before the governors of the Charity Schools.* These persons proceeded to prove that Whiston's publications put forward teachings against the established doctrine of the Church of England. Whiston was accused under 4 headings of publishing and spreading opinions in the University <u>contra religionem et statuta academiae.</u>

1. That the Father alone is the One God of the Christian religion, in opposition to the three divine persons, Father, Son, and Holy Ghost, being the One God of the Christian religion.

    This position is contrary to the first, second and fifth of the thirty-nine Articles, and to the Nicene and Athanasian Creeds.

2. That the creed commonly called the creed of St. Athanasius, is a gross and antichristian innovation and corruption of the primitive purity and simplicity of the Christian faith among us.

    This position is contrary to the rubric before the said creed, and the eighth article of religion.

3. That the canon of the scripture, the rule and guide of a Christian's faith, is that contained in the last of the ecclesiastical canons, ordinarily stiled Apostolical which all along appears to have been the standard of the primitive Church in this matter: I mean as including all books we now own for

---

*   The doxology which he used on this occasion, 25th Jan. 1705 at Trinity Church was as follows:

    "To whom, with the Father and the Holy Ghost, Three Persons and one God, be all honour, glory, thanksgiving, adoration and obedience rendered by us and all creatures, henceforth and forever more. Amen. (22)

canonical, and also the two Epistles of St. Clement, and the Constitutions of the Apostles of St. Clement, to which the Pastor of Hermas is to be added, as well as we have already added the Apocalypse of St. John.

4.   The doctrine of the Apostles appears to be a sacred book of the New Testament, long lost to the Christian World.

These two positions are contrary to the sixth of the thirty-nine Articles.

Mr. Whiston undertakes to prove clearly that the Apostolical Constitutions are the most sacred part of the canonical scriptures of the New Testament. Mr. Whiston asserts that the Doxology current in all these latter ages, Glory be to the Father, and to the Son, and to the Holy Ghost, is not the true Christian Doxology.

This position is against the Doxology received and established in the public liturgy. (23)

These books were proved to be dispersed by his directions in the University. Whiston defended himself against these accusations under eight headings. His defence attempted to show that his activities in these matters were not connected with the discharge of any public office in the University and that St. Clement's Church where he was charged with having spoken the doxology was outside the jurisdiction of the University. He also pointed out that the controversial work Sermons and Essays upon Several Subjects (1709) was printed in London.

In spite of his defence a sentence of banishment from the University was signed by Charles Roderick, the Vice Chancellor and eleven Heads of Colleges. His Professorship was likewise decreed void on October 25th 1711 by Dr. Lany, Master of Pembroke Hall, then Deputy Vice Chancellor and heads of eight colleges.(24)

Several attempts were made to bring Whiston before the

Court of Convocation for a condemnation of his heretical
views.[25]   Whiston published <u>An Account of the Convocations
Proceedings against Mr. Whiston</u> in 1711.   It seems that
although convocation agreed in condemning Whiston's views,
there was uncertainty about the authority required to enforce
a sentence of excommunication.

The procedures dragged on for several years and in
1712 the Convocation of Bishops and Archbishops of Canterbury
were still addressing the Queen to gain the necessary
authority.

> "May it therefore please your most gracious Majesty
> out of your known zeal for the honour of God and
> the good of his Church to lay this case before your
> reverend judges and others whom your Majesty shall
> in your Wisdom think fit for their opinion, how far
> the Convocation, as the law now stands may proceed
> in examining, censuring and condemning such tenets
> as are declared to be heresy by the laws of this
> realm, together with the authors and maintainors
> of them."   (26)

Whiston seems to have been quite unperturbed by all
these censures.   He was never in fact degraded nor excommunicated.   Whiston's lack of concern about the gravity
of the proceedings is demonstrated by the fact that while
waiting at the door of the Spiritual Court of St. Paul's,
he gave out his <u>Proposals for Finding out the Longitude at
Sea</u>.[27]

Dr. Sacheverell, Rector of St. Andrew's Church in
Holborn forcibly turned Whiston out of his church in 1721
with the result that Whiston left his house in Cross Street,
Hatton Garden, and moved to Great Russell Street in

Bloomsbury.*   Whiston's trial for heresy received so much public attention that it gave rise to a comic version known as <u>The Trial of William Whiston for defaming and denying the Holy Trinity before the Lord Chief Justice Reason</u> (1734).[28]

5.6   <u>NATURE OF WHISTON'S HERETICAL VIEWS</u>

In a letter written on April 5th 1711 to the Archbishop of Canterbury, Thomas Tenison (1694-1714), Whiston asks for the opportunity to justify his beliefs on the subject of the Trinity.   He disowns the charge that he is attempting to revive the heresy of Arius.

> "I do declare it was never my intention to assert the Arian heresy, strictly so called ... but ever confine myself to much ancienter and more authentic doctrines and languages of the scripture and the most primitive writers ...   In reality, I think no one ought to be at all led by any particular men, but to take their faith and practice from the most sacred and primitive writers, which lived long before the rise of the controversies of the fourth century;   as I have endeavoured to do in my <u>Account of the Primitive Faith of Christians</u>."   (29)

Whiston recalls the fact that many of his friends tried to restrain him from pursuing his study of Primitive Christianity as based on the Apostolical Constitutions. Dr. Richard Laughton, formerly his tutor at Cambridge,

---

* Henry Sacheverell (1674-1724) preached on the absolute necessity of God's chosen ruler, the Queen, and on the impossibility of resistance to royal authority in any form.   He violently disapproved of the toleration of dissenters whom he believed to be in league with the Pope.   Sacheverell was forbidden to preach for three years but he resumed his activities in 1713.   Due to his influence, every public figure was forced to re-examine his attitudes to the revolution of 1688.   Queen Anne presented Sacheverell to the rich living of St. Andrew's in Holborn.   Geoffrey Holmes in <u>The Trial of Dr. Sacheverell</u> (1973) gives a thorough analysis of the political and religious significance of this event.

Mr. Priest, Dr. Bentley and others all pleaded with him to recognise the error into which he had fallen.[30]

Whiston's five volumes Primitive Christianity Revived published in 1711 aroused a great deal of public interest in his claims for the Apostolical Constitutions. Accounts of Whiston's ejection from Cambridge and his trial before Convocation were also in print and the debate which followed involved Whiston in a pamphlet warfare with notable Churchmen and scholars.

In 1712, Samuel Clarke published his Scripture Doctrine of the Trinity. According to Clarke, unqualified worship is due to the Father alone and all reverence accorded to the Son or Holy Spirit must satisfy this primary condition. Clarke insisted that for the metaphysical Trinity of the creeds, he was substituting a Scriptural Trinity. He believed that the Athanasians were guilty of Sabellianism but the Arians and Socinians were also unsound. The strong rationalism which marked his position unquestionably favoured the trend towards Arianism and prominent Churchmen considered Clarke to be a very dangerous man.[31]

Whiston and Clarke maintained a close relationship from their first meeting in 1697 to Clarke's death in 1729. In 1730, Whiston published his Historical Memoirs of the Life of Dr. Samuel Clarke, in which he emphasises that as early as 1699, Clarke published his first theological work Three Practical Essays which dealt with the severity of the discipline of the Primitive Church. By 1705, Clarke had begun an investigation into the Primitive writers and had begun to suspect that the Athanasian doctrine of the Trinity was not the Doctrine of those early ages.[32]

In 1709, Whiston translated the Apostolical Constitutions into English.   Clarke undertook a revision and correction of this work.

> "We read a great part of it together as he
> corrected the rest by himself and sent me the
> corrections ...   he found about ten or twelve
> places which he hesitated about, he recommended
> me to go to our great and common friend Dr.
> Smalridge for the last correction of those most
> difficult places."   (33)

John Ernest Grabe (1666-1712) was one of Whiston's most scholarly opponents in the debate about the authenticity of the Apostolical Constitutions.   Grabe had left Prussia for England attracted by the fact that the apostolical succession was maintained without the 'superstitions of popery';   he was a doctor of divinity of Oxford University and his principal work was an edition of the Septuagint from the Alexandrian MS in the royal library. In 1711 Grabe wrote an attack on Whiston entitled An Essay upon two Arabic manuscripts of the Bodleian Library and that ancient book called the doctrine of the Apostles which is said to be extant in them;   wherein Mr. Whiston's mistakes about both are plainly proved.

Peter Allix accused Whiston of not giving sufficient notice to the studies already made by learned men of the Apostolical Constitutions.

> "... Mr. Whiston seemed not to me to have examined
> the marks of supposition which several learned men
> have found in the Apostolical Constitutions.   And
> indeed as learned from one of his friends, he had
> never read Dallaeus's book De Pseudepigraphis
> Apostolicis, where that learned man had demonstrated
> it to be spurious.   But according to Mr. Whiston
> that Book is the most canonical book in the whole

New Testament, because all the others are only
supported by its authority; but in the judgement
of the learned critics, both papists and protestants, it was forged by some impostor after
the first Council of Nicea." (34)

As a result of his examination of the writings of
Primitive Christianity, Whiston published his major work on
this subject in 1711, <u>Primitive Christianity Revived</u> in
five volumes. An outline of the five volumes is given here
to clarify the nature of the subject matter involved.

Volume 1.   An Account of the Convocation's proceedings
against me. A Dissertation on the Epistles
of Ignatius, with the Epistles themselves,
Greek and English and Eunomius's Apologetic.

Volume 2.   The Constitutions of the Holy Apostles,
Greek and English.

Volume 3.   A Vindication of those Constitutions.

Volume 4.   An Account of the Primitive Faith: with the
fourth book of Esdras, from the Latin and
Arabic.

Volume 5.   The recognition of Clement in English with
a preface and two appendices.

To all which is added a collection of small
tracts against Dr. Allix, Dr. Grabe,
Dr. Smalbroke.*

Whiston's <u>Letter to the Earl of Nottingham concerning
the Eternity of the Son of God and of the Holy Spirit</u>
published in 1719 quotes extensively from his <u>Primitive</u>

---

* Richard Smalbroke (1672-1749) was in 1731 nominated as
Bishop of Coventry and Lichfield. As Chaplain to Thomas
Tenison, Archbishop of Canterbury in 1711, he had written
several tracts denouncing Whiston's exposition of the
doctrine of the Trinity. The fact of such an eminent
prelate entering into controversy with Whiston only
served to highlight the problem of the doctrine of the
Trinity. Smalbroke's publications against Whiston
include <u>Reflections on Mr. Whiston's Conduct</u> (1711),
<u>The New Arian Reproved</u> (1711), <u>The Pretended Authority
of the Clementine Constitutions Confuted</u> (1714).

Christianity Revived.

In this letter, Whiston cites texts and testimonies:

(i)   For the original voluntary Generation and Creation
      of the Son of God before the World began and against
      his co-eternity with the Father.

(ii) which prove that the Holy Spirit was created or
      made, under the Father, by the Son, before the
      World began and against his Co-eternity. (35)

In 1721 Daniel Finch (second Earl of Nottingham)
published The Answer of the Earl of Nottingham to Mr.Whiston's
letter to him concerning the eternity of the Son of God.*
This publication gave rise to others centred on the Divinity
of Christ such as that by Edward Welshman in 1721, A conference with an Arian occasioned by Mr. Whiston's reply to
the Right Honourable the Earl of Nottingham.

Whiston took pains to prove the authenticity of the
Apostolical Constitutions by appealing to early Christian
authors such as St. Clement and St. Irenaeus who had given
them unqualified approval.

In 1715 Whiston published St. Clement's and St.
Irenaeus's Vindication of the Apostolical Constitutions.

      "... I shall beg leave to produce a most authentic
      witness or record to the contrary (i.e. to prove

---

* Finch had been appointed secretary of state at the accession of Anne in 1702. Throughout the trial of Sacheverell
in 1710, he took Sacheverell's side. After his retire-
ment from office, Finch lived at Burley-on-the-Hill near
Oakham, Rutland. In his 1721 publication, Finch answers
Whiston's 1719 letter and takes pains to show that the
Eternity of the Son of God and of the Holy Spirit are
proved by Scripture and tradition and accuses Whiston of
misrepresenting the Christian faith.
      "The Jesuits in China have formed a new gospel
      representing our Saviour only in his most exalted
      state, concealing his Crucifixion: on the other
      hand, you [Whiston] are depressing him into a
      creature, not concealing but denying His Divinity."(36)

that the <u>Apostolical Constitutions</u> are authentic)
... I mean, the known epistle of the Church of
Rome, or Clement its Bishop, to the Corinthians;
of that very Clement, whose name St. Paul tells
us, was in the book of Life; and by whose hand
and attestation all the eight books of the
Apostolical Constitutions are in all the manuscripts recommended to the Churches as really
canonical books of the New Testament." (37)

By the year 1736, Whiston had been responsible for
many publications which examined the Arian heresy and
concluded by condemning Athanasius as a heretic. An
anonymous 'lover of truth and of the true religion' published a collection of twenty-two articles entitled
<u>Athanasian Forgeries, Impositions and Interpretations</u>
<u>collected chiefly out of Mr. Whiston's Writings</u>. In a
later publication, Whiston admitted that he himself was the
author of this work.

In 1741, when Whiston was seventy-four years of age,
he published <u>An Appeal to Thirty Primitive Councils against</u>
<u>the Athanasian Heresy</u>, divided into thirty-six headings
covering councils between the years 265 and 360. He
concludes:

"... we shall find that before the end of the year
361, on the Athanasian side, not one pretended
General Council; not one uncontested Great
Council and indeed no more than three or four
small councils were Athanasian." (38)

In 1740, Whiston published a work which reflected his
doubts about the eternity of the punishments of Hell:
<u>The Eternity of Hell Considered: or a Collection of Texts</u>
<u>of Scripture and Testimonies of the first three centuries</u>
<u>relating to them</u>. His rejection of eternal torment is
based almost solely on scriptural and early patristic

authority.   Whiston claims that no Christian writer earlier
than Tertullian asserted the doctrine of the true eternity
of rewards and punishments.[39]   In his Astronomical
Principles of Religion, Natural and Revealed (1717), Whiston
makes a synthesis of his views on natural philosophy and
his scriptural interpretations.   As already mentioned in
the chapter dealing with his cyclic cosmology, Whiston
identifies the conflagration caused by the approach of the
comet with the place of torment described in Scripture.[40]

For further detailed information about Whiston's
heretical works, an Appendix to this chapter gives a short
account of the content of each of his religious works.

5.7   THE SOCIETY FOR PROMOTING PRIMITIVE CHRISTIANITY
AND THE PRACTICE OF INFANT BAPTISM

Whiston was the founder of A Society for Promoting
Primitive Christianity which met regularly for two years,
1715-1717.   The Society, a group of twelve persons, was to
meet every Friday at five o'clock at Whiston's House in
Cross Street, Hatton Gardens, where a Primitive Library was
to be housed.   At the end of Whiston's 1712 publication on
Infant Baptism is an advertisement relating to the Library
at Cross Street, Hatton Gardens.

ADVERTISEMENT

In Cross Street, Hatton Garden, is collecting a
Primitive Library to contain all the Sacred and
Primitive books of the first four centuries;
with such as are immediately necessary for the
understanding of them, and no other.   All that pay
the owner 10s. at first and 5s. a year afterwards,
may use it, and borrow out of it at pleasure. (41)

The Minute Book of the Society, a large parchment-bound volume, survives in the Bodleian. The first three chairmen were Dr. John Gale, champion of the Anabaptists, Arthur Onslow and Thomas Emlyn. At the meeting of October 14th 1715, it was decided to examine

> "the antiquity and genuineness of two Epistles still extant and ascribed to Clement of Rome as written by him to the Corinthians in the first century and published from the Alexandrian manuscript and in special concerning the smaller epistle now commonly called his second epistle." (42)

The meeting on October 21st 1715, centred upon two detailed tables referring to citations by several authors of sections of the larger and smaller Epistles of Ignatius. One hundred citations from twenty-nine authors were recorded for the Larger Epistles of Ignatius while eighty-two citations from ten authors were recorded for the smaller epistles. The conclusion of the meeting seemed to be,

> "that the internal characters were arguments for the genuineness of these epistles and were not contradicted by any against it; and that the external testimonies though later and weaker, were yet on the same side; and that therefore the positive arguments were for the genuineness of the same epistles: to which opinion therefore the Society unanimously inclined." (43)

The last meeting of the Society took place on 28th June 1717.

Whiston's uncle, Joseph Whiston, had published several works on the practice of infant baptism. Infant-baptism from heaven, and not of men (1670); An essaye to revive the primitive doctrine and practice of Infant-baptism, in the resolution of four questions (1676); The right method for the proving of Infant-Baptism, with some reflections on some late tracts against Infant-baptism (1690).

In 1712, Whiston published a work on Infant baptism whose full title is explanatory of the contents. _Primitive Infant-Baptism Revived or an Account of the Doctrine and Practice of the Two First Centuries concerning the Baptism of Infants; in the words of the Sacred and Primitive Writers themselves_. There is reason to believe that Whiston's work of 1712 leans heavily on the earlier writings of Joseph Whiston. The view expressed in the last-mentioned work is summed up in a section entitled:

"Observations From the Whole

1. That the Baptism of such Infants as are capable of Catechetick Instruction, and have been under it about three years, is for certain the Law of the Apostolical Constitutions, and of the known books of the New Testament.

2. That the Baptism of Infants, or such as are too young to be capable of the same instructions is utterly groundless.

. . . . . .

9. The use of Dipping, and even of Trine Immersion itself is plainly sacred and unalterable in Christian Baptism; and the later methods of pouring on or sprinkling water only, utterly unjustifiable among Christians." (44)

Whiston bases most of his arguments on texts from the _Apostolical Constitutions_. He urges that women should take part in the ministry of the church.

"Neither grown women, nor girls can be decently baptised as Christ has appointed, till the sacred order of _Deaconesses_ is revived in the Church." (45)

Whiston believed that Confirmation is only a part of Baptism, or the entire solemnity of initiation. In 1713,

Whiston published The Liturgy of the Church of England reduced to near the Primitive Standard. This work includes the form to be observed for the Ministration of Baptism and Confirmation.[46]

Whiston did actually administer the sacraments according to this rite. Several instances are recorded in his Commonplace book.

> "Feb. 21 1714.
> ... being the Lord's Day, I baptized Mr. John and Mrs. Elizabeth Shelswell, with the trine immersion; according to the form published by myself in my Revised Liturgy of the Church; with the office of Confirmation and Eucharist at the same time. (present about 17). [47]

Mr. Shelswell is also mentioned as being present at the first meeting of the Society for promoting Primitive Christianity.

Sarah Barker, daughter of Sarah Whiston and Samuel Barker was born on 21st May 1717. The document records that,

> "her baptism was deferred till she should grow up."[48]

5.8   WHISTON'S PRIMITIVE CATECHISM AND HIS WORK FOR THE CHARITY SCHOOLS

Whiston was associated with the setting up of charity schools at Cambridge for the education of about three hundred poor children. He preached the sermon to commemorate the inauguration of this work at Trinity Church on 25th January 1705. In February of the same year he wrote An Account of the Charity Schools Lately erected at Cambridge. These schools were designed to,

"... make some provision for the instruction of those
children the poverty of whose parents made it very
difficult if not impossible for themselves to provide for their education."

Whiston describes the origin of these schools as

follows:

"...   About the middle of the Year 1703, it was
proposed to several ministers of the town of
Cambridge to make an attempt for the erecting of
Charity Schools; and upon their ready and
cheerful consent, the design was immediately
communicated to some of the heads of the
University and particularly soon after to the
Right Reverend the Bishop for their approbation,
direction and assistance.

.....

... by Christmas that year (1703), they found
themselves enabled actually to provide for the
education of about 260 poor children ...   Since
that time the contributions and with them the
number of poor children are increased; so that
at present three hundred poor children, even most
of those whose parents are really unable to pay
for their education, are now taken into these
schools and carefully and religiously brought
up there."   (49)

The pamphlet goes on to give an account of the

Agreement for the Erecting of Charity Schools in Cambridge.

The fifth item listed was:

"V   When any of us in our turns shall be chosen
stewards, we will catechise the children once
every month, (Namely, the last Thursday in the
month at one o'clock). We will inspect the
schoolmasters and mistresses in their employment; and see how the scholars improve and
behave themselves."

Whiston was appointed as steward for a time and

records:

"I confess that my monthly day of catechising about
ninety of them, when I was their steward, seemed
to me to be the best spent day of the whole
month." (50)

In 1709, Whiston was obliged to give up his monthly lecture to the Charity schools because his heretical views had become known.   Bishop Patrick of Ely was urged by the minister of St. Clement's parish Cambridge to charge Whiston,

> "that in his explanations of the Church catechism
> he advanced notions contrary to those of the
> church and also that once when he read the
> prayers aloud, he omitted the third and fourth
> petitions of the Liturgy."   (51)

The Rules and Orders for the Charity Schools at Cambridge are listed.   The hours appointed for teaching were:

> "from seven to eleven in the morning, and from one
> to five in the evening, the summer half year;   and
> from eight to eleven in the morning and from one
> to four in the evening, the winter half year."

The boys were to be taught, "to read, write and cast account" while the girls were to be taught "to read, to write and work."

Great emphasis was placed on teaching the catechism. The prayers and graces which they were to learn were exactly specified.   Those children below ten years of age were to learn the Church Catechism and those above ten years of age were to learn Mr. Lewis's Explanation of the Church Catechism.   The efficacy of rewards during this rather formalised instruction was tested.

> "The steward then as soon as the children are
> seated in order, the boys in one place and the
> girls in another ... examines every one of them
> in the catechism;   and if the answers be to his
> satisfaction, gives them a penny or half-penny
> a-piece for their encouragement....   And I shall
> take leave to say that small sums spent in that
> manner are found to have a great effect in the
> improvement of the children."

In 1718, Daniel Whiston, youngest brother of William, published <u>A Primitive Catechism</u>. Whiston revised and modified this work in the same year.

> "... I myself greatly approved and improved [<u>The Primitive Catechism</u>] and I published [it] under the title of a Presbyter of the Church of England, and still insert it among the catalogue of my own writings." (52)

It was designed for use in the Charity schools, but by this time Whiston was not in a position to introduce it into the Charity schools at Cambridge. Towards the end of his life, Whiston did support a small Charity school in which he used this <u>Primitive Catechism</u> which relies heavily on the <u>Apostolical Constitutions</u>,

> "It is this <u>Primitive Catechism</u> ... which I ever make use of for the instruction before Baptism in such as have not yet been baptised ... Accordingly, when I, about seven years ago (about the year 1740) supported a charity school of ten boys and ten girls by my own and some friends contributions, for two years and a half, I went, at least one day every week to hear them repeat, and explain to them the Epitome of this Primitive Catechism." (53)

The full title of this work was <u>A Primitive Catechism by way of Question and Answer in two Parts   Useful for Charity Schools</u>.* The first part of the work gives texts of scripture for the proof of the several answers. The second part omits the scripture texts and is probably the Epitome mentioned by Whiston in the extract just cited.

---

* A second enlarged edition of this catechism was printed in 1751. The British Museum copy of this work has manuscript notes and additions by Whiston. (54)

5.9 CORRESPONDENTS OF WHISTON REGARDING PRIMITIVE CHRISTIANITY

Correspondence survives between Whiston and Robert Fleming who is known principally for his three volume work entitled Christology. One extended letter from Fleming and three short replies from Whiston all date from the year 1709. Fleming challenges Whiston about his intention to publish a treatise on the Trinity based on Arian principles. Fleming points out that his presentation of the Trinity is also based on scriptural enquiries.

> "... Some that know you inform me of your intention to publish a treatise on the Trinity upon Arian principles. If so, allow me to give you some short hints of my own sentiments on this great head. ... And yet in my scriptural inquiry into the divinity as well as the pre-existence of Christ, I was far from founding these things on anything but scripture as you will find if you care to look into Christology Book 2, Chapters 4 and 5." (55)

Whiston is at pains to defend the authenticity of the Apostolical Constitutions and reproaches Fleming for not having adequately examined the claims of these documents.

> "... And you I perceive, have never had the assistance of the Constitutions of the Apostles by St. Clement which are and ever were a part of the Canon of the New Testament."
>   6th Oct. 1709 (56)

> "... I and a friend with me have gone much deeper in the search about the Constitutions of the Apostles than any of those authors you mention. My papers are now in Dr. Grabe's hands. When you see them, you will find how little any of the moderns have understood the matter."
>   19th Oct. 1709 (57)

> "... You take that canon of the sacred books which the Church of Rome has transmitted to you, I take the original one, attested to by all Antiquity." (58)

Another correspondent of Whiston was John Jackson,
Master of Wigston's Hospital in Leicester (1686-1768), a
zealous advocate of Arianism who wrote some tracts against
the doctrine of the Trinity but whose best known work is
Chronological Antiquities (3 volumes, 1752).*

In 1716, Jackson expresses his interest in Whiston's
work on Primitive Christianity.

> "... Thanks for your honest writings, Primitive
> Christianity Revived, and particularly to tell
> you that I verily think you have proved the larger
> epistles of Ignatius, to be the true genuine ones,
> and they are of great importance in the cause of
> truth.   I am one who upon all occasions vindicate
> you from the clamour of unreasonable men of the
> brotherhood, and I hope to live to see the pro-
> fession of true christianity flourish and be
> rewarded."   (59)

The next surviving letter between Jackson and Whiston
written on 17th April 1717 extends to six closely written
foolscap sheets.   In this document, Jackson lays out in
detail his reasons for not accepting the Apostolical
Constitutions as authentic Apostolical writings to be
numbered among the sacred books.   Jackson's arguments are
carefully reasoned and throughout he shows respect for
Whiston's judgements while choosing to differ in his own
conclusions.   Jackson makes the point that this work was
never quoted by any writer of the first three centuries

> "... I do not find that any such book or body of
> written doctrine and laws as the Constitutions
> pretend to be, is ever cited or mentioned by any
> writer of the three first centuries; or reckoned
> as part of the canon of Scripture, in any age for
> a thousand years, as Dr. Whitby has observed;

---

* John Jackson (1686-1768) was rector of Rossington,
York and prebendary of Wherwell in Southampton.

which observations are, I think, claims sufficient
to overthrow their sacred character." (60)

In the same letter, Jackson discusses the question of
Infant-Baptism. He does not agree that catechesis always
preceded Baptism in the early Church.

Whiston was not alone in his efforts to revive a
primitive form of Christianity. In 1724, he refers to a
young man, Mr. William Paul, a student at the University of
Glasgow who was labouring,

"for the restoration of Christian Liberty and
primitive Christianity." (61)

Thomas Emlyn, another close associate of Whiston
(their relationship dating from Whiston's time at Lowestoft
and Kessingland) also worked for the promotion of Primitive
Christianity in Ireland. Emlyn was a native of Stamford
in Lincolnshire and was educated at an academy of non-
conformists in Leicestershire. Emlyn's works were published in two volumes in 1746.(62)

In December 1710 Whiston felt himself obliged to
dissociate himself from further participation in The Society
for Promoting Christian Knowledge founded by Thomas Bray
(1656-1730). This society thought itself only capable of
supporting things as they then stood in the Church of
England as by law established.(63)

5.10 VIEWS OF MODERN BIBLICAL SCHOLARSHIP REGARDING THE APOSTOLICAL CONSTITUTIONS

The Apostolical Constitutions comprise a collection of
ecclesiastical regulations in eight books, the last of which
concludes with the eighty-five canons of the Holy Apostles.
By title the Constitutions profess to have been drawn up by
the Apostles and to have been transmitted to the Church by
Clement of Rome.   The Constitutions as a whole remained
unknown in the West until they were published in 1563 by
the Jesuit, Turrianus.   At first they were received with
enthusiasm but their authenticity soon came to be impugned
and their true significance was largely lost.

The Apostolical Constitutions are the last stage of a
process of compilation and crystallisation of unwritten
Church custom.   By separating off the sources used by the
author from his own additions, it becomes clear that the
Constitutions are the work of one man.   His polemic is
addressed against the dying heresies of the third century
with an inclination towards Arianism.   The conclusion seems
to be that the author was a semi-Arian of the East belonging
perhaps to the circle of Lucian of Antioch.   The compiler
then seems to have been a Syrian who also wrote the spurious
Ignatius epistles.   Various dates are assigned for the
compilation, most pointing to 340-380 A.D.[64]

Whiston was one of several scholars who became interested in the authenticity of the Apostolical Constitutions.
His claims on its behalf as the most sacred part of
Scripture were certainly exaggerated.   His studies were,
however, part of a great revival of interest in the practice

of early Christianity.   Controversial arguments about the
authenticity of documents were a necessary phase in the
recognition of a hierarchy of value in the written legacy
of the early church.

CHAPTER 5 - REFERENCES

1. R.C.H.M. Bath MS. Vol. I, p.311.
               Edward Young to the Duchess of Portland.

2. Whiston, W.
               Memoirs of the Life and Writings of Mr.
               William Whiston (1749), p.30.

3. Cragg, G.R.
               From Puritanism to the Age of Reason.

4. Locke, John.
               Essay Concerning Human Understanding (1690)
               Book IV, Ch. XVIII, Sect. 4.

5. Willey, Basil.
               The Eighteenth Century Background (1972).
               Pelican Edition, p.16-17.

6. Op. cit. (reference 2), p.130.

7. B.M. Add. MS. 24197.

8. Op. cit. (reference 2), p.333.

9. R.C.H.M. Egmont MS. 1, p.288.
               Diary of the First Earl of Egmont.

10. Prideaux, H.
               The Old and New Testament connected in the
               History of the Jews and Neighbouring
               Nations (1718).   B.M. copy. Shelf No. 873,
               1. 11, p.26.

11. Whiston, W.
               Authentic Records belonging to the Old and
               New Testament (1727, 1728).   B.M. copy.
               Shelf No. 873, 1. 6 and 873, 1. 7.

12. Op. cit. (reference 2), p.363.

13. Bodleian MS.   Eng. Theo. C. 60.

14. Encyclopedia Britannica, Eleventh Edition, Vol. 15,
               p. 517.

15. Field MS. F 137.   Jackson to Whiston, October 16th,
               1742.

16. Op. cit. (reference 10).

17. Birch, Thomas.
               The Life of the Honourable Robert Boyle
               (1744). p.353-354.

18. <u>Boyle Lectures</u>, Vol. 2 (1739), p.320.

19. Ibid., p.329.

20. Kelly, J.N.
    <u>Early Christian Creeds</u> (1950).

21. <u>Biographia Britannica</u> (1766), p.4205.

22. Whiston, W.
    <u>A Sermon preached at Trinity Church</u>
    C.U.L. Cam. d. 705 (2).

23. Op. cit. (reference 21), p.4206.

24. Documents relating to Whiston's Banishment from
    Cambridge. The University Archives,
    Old Schools, Cambridge. C.U.R. 39.8
    Numbers 6, 7, 9.

25. Bodleian MS. Tanner 282. F 150, 151, 152, 155, 156.

26. Ibid.   F 151, 152.

27. Op. cit. (reference 21), p.4209.

28. Gordon, T. <u>The Trial of William Whiston for defaming</u>
    <u>and denying the Holy Trinity before The</u>
    <u>Lord Chief Justice Reason</u> (1734).
    Fellows Library, Clare College.

29. Whiston, W.
    <u>An Account of the Convocation's proceedings</u>
    <u>with relation to Mr. Whiston</u> (1711),
    p.13-14.

30. Op. cit. (reference 2), p.151.

31. Cragg, G.R.
    <u>Reason and Authority in the Eighteenth</u>
    <u>Century</u> (1964). C.U.P.

32. Whiston, W.
    <u>Historical Memoirs of the Life of Dr.</u>
    <u>Samuel Clarke</u> 3rd edition (1748), p.7.

33. Ibid.   p.16.

34. Allix, P. <u>Remarks upon some places in Mr. Whiston's</u>
    <u>Books</u>, Second edition (1711), p.9.

35. Whiston, W.
    <u>Letter to the Earl of Nottingham</u> (1719),
    pp. 9 and 34.

36.   Finch, D.   The Answer of the Earl of Nottingham to
                  Mr. Whiston's Letter (1721).   Sixth
                  edition, p.79.

37.   Whiston, W.
                  St. Clement's and St. Irenaeus' Vindication
                  of the Apostolical Constitutions
                  Second edition (1716), p.12.

38.   Whiston, W.
                  Appeal to the Thirty Primitive Councils
                  against the Athanasian Heresy (1741), p.16.

39.   Walker, D.P.
                  The Decline of Hell.   Routledge, Kegan
                  Paul (1963).

40.   Whiston, W.
                  Astronomical Principles of Religion   (1717)
                  p.156.

41.   Op. cit. (reference 13).

42.   Ibid.

43.   Ibid.

44.   Whiston, W.
                  Primitive Infant Baptism Revived (1712),
                  p.44-46.

45.   Ibid.   p. 45.

46.   Whiston, W.
                  The Liturgy of the Church of England reduc-
                  ed to the Primitive Standard.   Second
                  edition (1721).

47.   Bodleian MS. Eng. misc. d. 297 F.7.

48.   Ibid.

49.   Whiston, W.
                  An Account of the Charity Schools lately
                  erected at Cambridge (1705), p.16.

50.   Op. cit. (reference 2), p.133.

51.   Op. cit. (reference 21), p.4206.

52.   Op. cit. (reference 2), p.13.

53.   Ibid.   p. 288.

54.   Whiston, W.   A Primitive Catechism (1751).
                  B.M. Copy. Shelf No. 874, 1.21.

55.   Bodleian MS. Rawlinson 105, F.29.

56.   Ibid. F.32.

57.   Ibid. F.38.

58.   Ibid. F.44.

59.   Field MS. F.105.   Jackson to Whiston, October 31st
                                                    1716.

60.   Ibid. F.106. Jackson to Whiston, April 17th, 1717.

61.   Op. cit. (reference 2), p.310.

62.   Reeve, R. Manuscript (Volume 4) for the history of
                 Mutford and Lothingland, p.77.
                 Ipswich County Record Office.

63.   Field MS. F.99   W. Whiston to the Society for
                 Promoting Christian Knowledge, December 1710.

64.   Op. cit. (reference 14), p.199-201.

APPENDIX

CHRONOLOGICAL TABLE OF THE SIGNIFICANT RELIGIOUS
PUBLICATIONS OF WILLIAM WHISTON

This Appendix sets out the subject matter of each work and
to explain any contributory factors which led to its
publication.

In those cases where the work is composite in nature, consisting of prefaces, appendices and supplements this table
shows the subdivisions of such works listed under one title.

1702

A Short View of the Chronology of the Old Testament and of the Harmony of the Four Evangelists (1702)

Whiston dedicated this work to John Moore, Bishop of

Norwich, whose chaplain he had been until 1702.   In the

Epistle Dedicatory, Whiston gives due praise to the work of

Archbishop Ussher, Annales Veteri et Novi Testamenti (1650-

1654).   Ussher's work on New Testament Chronology is based

on Bishop Richardson's Harmony of the Four Evangelists.

Whiston claims to have made further adjustments and improve-

ments on Ussher's work so that his version of the Chron-

ology and Harmony are not opposed to any authentic existing

evidence.   It begins with four pages of tables of the Kings

of Judah and Israel.   The chronology, which forms the first

part of the book, is an attempt to link the biblical time-

scale to that of sacred and profane authors.   Whiston also

claims that the Chinese chronology agrees exactly with that

drawn from the Hebrew text of the Old Testament.

The Short View of the Harmony of the Four Evangelists

forms the second part of this work.   The first section con-

sists of twenty propositions which clarify key events or

interpretations of the work of the Evangelists. The Harmony itself is printed in pages, each of four columns for the evangelists. The work concludes with a summary of the Harmony in table form, giving references to gospels and dating the events according to the calendars of the Jewish and Christian era.

The B.M. copy of this work (Shelf Number 873, 1.5) contains many manuscript notes and corrections which Whiston made in preparation for a second edition of this work.

1706

An Essay on the Revelation of Saint John (1706)

This essay is divided into three parts. Part I comprises a series of eighteen propositions which attempt to elucidate some of the incidents in the Revelation of Saint John and place them against the time scale of Israel's history. There follows a chart which attempts in pictorial form to represent these divisions of time over the 2,000 years before the birth of Christ. Part II of the essay is concerned with the prophecies contained in the Sealed Book while Part III deals with the prophecies contained in the Open Codicil. Whiston interprets some of these prophecies as literally fulfilled in civil and ecclesiastical history. Bound with this essay are two short dissertations on Mark II and Matthew XXIV. There follows a collection of several hundred Scripture Prophecies relating to the times after the coming of the Messias.

This work formed a basis for the Boyle lectures which Whiston delivered in 1707.

1708

The Accomplishment of Scripture Prophecies (1708)

These were the eight sermons which Whiston preached at St. Paul's Church in the year 1707 as the Boyle lectures of that year. The content of these sermons is discussed in Chapter five. To this work is added a dissertation to prove Our Blessed Saviour ascended up to Heaven the evening of that very day on which he rose from the dead. This short essay contains some interesting links between Whiston's cosmological and theological ideas. Whiston defines Heaven as an EXPANSUM or open place, on high, more clear and light than the regions below. Whiston identifies heaven with the upper regions of the air. He puts forward the strange interpretation that Our Lord ascended and descended more than once during the forty days between his Resurrection and Ascension, the depth of the atmosphere being only 54 or 55 miles from the earth. Thus Christ was spared the immense journey of millions of miles to the limit of the astronomical universe!

1709

Sermons and Essays upon Several Subjects (1709)

This set of ten sermons and essays were written during the period 1699 to 1708. To these is added the Latin version of Novatian's De Trinitate. By 1709 Whiston had developed his interest in the revival of Primitive Christianity and there is evidence of this in Essay X, Advice for the Study of Divinity. Whiston advises the student of Divinity that the canon of the Scripture, the rule

and guide of the Christian's Faith and Practice is that contained in the last of the ecclesiastical canons, namely the
Apostolical Constitutions.

Sermon V of this collection was preached at Trinity
Church in Cambridge on January 25th 1705.   Whiston used
his own doxology at the conclusion of this address to the
governors of the Charity Schools.   This heretical formulation was later to be one of the causes of his ejection
from the University at Cambridge.   The British Museum copy
of this work contains numerous manuscript notes made at
later periods of his life. (Shelf Number 873, 1.15.)

1710

An Essay upon the Epistles of Ignatius

In this short work of 46 pages dated Feb. 29th 1710,
Cambridge, Whiston sets out a general Proposition:
The larger epistles of Ignatius, which of late have been
stiled his interpolated epistles are alone the genuine and
original epistles of that father.   The small epistles are
only an epitome of several of the larger, made about or
after, the fourth century of the Church.   Whiston extends
the number of the authentic epistles of Ignatius from seven
to ten.

Most scholars recognise only the first seven epistles
as authentic documents.

1711 and 1712

Primitive Christianity Revived (1711)

This represents Whiston's major work in his efforts to

revive the practice and doctrine of the early Church.   It
was written after his expulsion from the University at
Cambridge and is defiantly bold in its dedication to
royalty, church leadership and academic institutions.

The dedications are as follows:-

Volume 1.   To John, the Archbishop of York and to the
            Bishops and Clergy of the Lower House of
            Convocation.

Volume 2.   To Her Majesty Queen Anne.

Volume 3.   To the University of Oxford.

Volume 4.   To the University of Cambridge.

The whole is dedicated to Thomas Tenison, Archbishop of
Canterbury and to the Bishops and Clergy of the Lower House
of Convocation.

Volume 1 is concerned with The Epistles of Ignatius,
Bishop of Antioch, both the larger and the smaller in Greek
and in English, with various readings from all the Greek
Manuscripts.

The Historical Preface is preceded by Eunomius's
Apologetick (pages 1-30).   Basil the Great firmly upheld
the Nicene doctrine of the Trinity and wrote a confutation
of the teaching of Eunomius.

The Historical Preface to Volume 1 is a composite work
and the following items are included, (pages i-cxxvi).
In these pages Whiston sets forth the history of the events
and influences which have brought him to his present state
regarding the doctrine of the Trinity and his attempts to
revive a primitive form of Christianity.   The major part
of this Historical Preface is devoted to the printing of
correspondence between Whiston, Church authorities,

authorities at the University in Cambridge, friends and acquaintances, regarding his tenets concerning the doctrine of the Trinity and the authenticity of the Apostolical Constitutions.

The first Appendix to the Historical Preface, pages cxxxvii-clxxv) gives an Account of the Authors Prosecution at, and Banishment From the University of Cambridge. Whiston goes to great pains to document his trial and to show that the condemnation was unjust.

There follows Appendix II, pages 1-61, which is An Account of the Convocation's Proceedings with relation to Mr. Whiston. Pages 63-69 form A Supplement to the Account of the Convocation's Proceedings with relation to Mr. Whiston. These few pages give the judgement of the Archbishop and Bishops and Clergy of the Province of Canterbury concerning the assertions contained in Whiston's works. It was on the basis of statements which Whiston printed in his Historical Preface to Primitive Christianity Revived that Whiston's views were condemned as heretical.

In a Postscript Whiston bitterly laments what he considers to be a series of gross injustices, which he has suffered because of his personal views.

The Greek and English texts of the Epistles of Ignatius are preceded by a Dissertation upon the Epistles of Ignatius (pages 1-101). Whiston maintains that the larger Epistles of Ignatius are genuine while the smaller epistles date only from the middle of the fourth century. The English Translation of the larger Epistles of Ignatius is due to Whiston while the translation of the smaller epistle is due to the Bishop of Lincoln.

The Larger and Smaller Epistles of Ignatius are followed by three more Epistles of Ignatius, Bishop of Antioch which Whiston also believes to be authentic:-

        To the inhabitants of Tarsus
        To the Antiochians
        To Hero the Deacon.

These three epistles are not generally considered to be authentic.

Volume II consists of the eight books of the Constitutions of the Holy Apostles by Clement in Greek and English.   The English version of the Constitutions is by Whiston.   Samuel Clarke assisted him in the translation.

Volume III comprises Whiston's Essay on the Apostolical Consitutions in which he sets out to prove that they are the most sacred part of the Canonical Books of the New Testament.

Volume IV gives An Account of the Faith of the Two First Centuries.   In twenty-three articles, Whiston gives his version of the doctrine of the Trinity and of the Incarnation, taking as his authorities the books of the New Testament and the most primitive writers to about the year 190 A.D.   The works held as authoritative include the Constitutions of the Apostles and the Larger Epistles of Ignatius of Antioch.

An Appendix to Volume IV deals with the Primitive Doxologies (pages 1-56).   Then follows a translation from the Arabic of the Second Book of Esdras.   The translation is by Mr. Ockley.

The 1711-1712 edition of Primitive Christianity Revived

was in five volumes, the fifth volume being a collection of seven small tracts of which Whiston was the author.   These were available separately but Whiston published them in one volume as they all bore close relationship to the subject matter of his Primitive Christianity Revived.   Six of the seven tracts were answers to his critics, Dr. Peter Allix, Mr. Chishull, Dr. E. Grabe, Dr. Richard Smalbroke and Mr. Thirlby.

A second edition of Primitive Christianity Revived appeared in 1712 in four parts which were evidently bound into one large volume.

The British Museum holds five sets (24 volumes) of Primitive Christianity Revived which belonged either to Whiston himself or to his son-in-law, Samuel Barker (Shelf Numbers 873 m. 1-24).   Five volumes of Primitive Christianity Revived are interesting in that they are printed with interleaved plain pages on which some person has started to make a Latin translation of the work in manuscript.   The Latin copy is complete as far as page 102 in Volume 2 and has been corrected by Whiston.   (Shelf Number 873 m. 10-14).

1712

A Collection of Small Tracts formerly Published (1712)

In a note to the reader, Whiston explains that although these seven tracts have already been published separately, they are not to be found in any of the five volumes of Primitive Christianity Revived.   For the convenience of the reader and the purchaser of the 5 volumes mentioned above, they are now available as a set.   The seven tracts, all written by Whiston are as follows:-

1. <u>A Reply to Dr. Allix's Remarks on some places of Mr. Whiston's Books either Printed or in Manuscript</u> (1711)

An Appendix to this has three components

   (a) <u>The Preface to the Doctrine of the Apostles</u>

   (b) <u>Propositions containing the Primitive faith of Christians about the Trinity and Incarnation</u>

   (c) <u>A Letter to the most Reverend Archbishop of Canterbury, President of Convocation.</u>

2. <u>A Second Reply to Dr. Allix</u> (1711)

This has two postscripts

   (a) To Mr. Chishull on the occasion of his visitation sermon.

   (b) Comments on an anonymous pamphlet recently published entitled, <u>Reflections on Mr. Whiston's Conduct.</u> (The author of these Reflections was Richard Smalbroke).

3. <u>A Reply to the Considerations on the Historical Preface and the Premonition to the Reader</u> (1711)

4. <u>Remarks on Dr. Grabe's Essay upon two Arabic Manuscripts of Bodleian Library</u>, Second edition (1711)

5. <u>Animadversions on a Late Pamphlet entitled The New Arian Reproved</u> (1711) (The author of the <u>New Arian Reproved</u> was Richard Smalbroke).

6. <u>Athanasius Convicted of Forgery in a letter to Mr. Thirlby of Jesus College in Cambridge</u> (1712)

7. <u>Primitive Infant Baptism Revived</u> (1712)

The British Museum copy of this Collection of Tracts (Shelf Number 873. 1.1) has many manuscript notes and additions.

<u>1712</u>

<u>Athanasius Convicted of Forgery</u> (1712)

   This work was in the form of a letter to Mr. Thirlby of Jesus College, Cambridge, who in 1712 had written a pamphlet entitled <u>An Answer to Mr. Whiston's Seventeen</u>

Suspicions concerning Athanasius in his Historical Preface.

In this 1712 work, Whiston sets out to show that his suspicions against Athanasius have now hardened into a certain account of his forgery. In his Memoirs, Whiston acknowledges the substance of the views which he expresses in this work are derived from Dr. Robert Cannon and his friend Richard Allin of Sidney Sussex College, Cambridge.

1712

The Old Paths Restored

In 1712, Whiston published in the form of a single sheet a work printed at Boston, New England and dedicated to Dr. John Edwards (1637-1716). Edwards had been a fellow of St. John's College Cambridge but his position had been untenable because of his Calvinistic views.

This short pamphlet, Old Paths Restored, was intended as a

> "brief demonstration that the doctrine of grace
> hitherto prescribed in the churches of the non-
> Conformists are not only asserted in the sacred
> scriptures but also in the articles and
> homilies of the Church of England."

Whiston states that he does not himself believe the main part of the doctrine expressed in this pamphlet but he thinks that salutary lessons may be learnt about the source of divisions in the Church.

1713

Reflections on an Anonymous Pamphlet entitled, A Discourse of Free Thinking (1713)

This work was one of many which appeared as a condemnation of the prominent Deist writer, Anthony Collins, who published A Discourse of Free Thinking in 1712.

Collins' work was calculated to raise doubts whether the tradition of the early Church and the revelation of God's word in Scripture were compatible with a rational approach to Christianity.

Whiston questions the readiness of the Deists to abandon reliance on tradition and scripture simply because occasional anomalies can be detected. He used the occasion to express concern about the present state of the corruption of the Church and particularly the scandalous behaviour of several prominent clergymen. This work appeared in three editions in 1713. The British Museum copy of the third edition of this work (Shelf Number 873. 1.1(3)) contains many manuscript notes and corrections by the author.

1713

The Liturgy of the Church of England reduced nearer to the Primitive Standard (1713)

This work was first published in 1713. A second edition with corrections was printed in 1750 and is found bound in with Volume III of his Memoirs. The contents include forms of morning and evening prayer, order for the Administration of the Lord's Supper, the ministration of Baptism and Confirmation, the order for the burial of the Dead and other prayers.

1713

The Christian's Rule of Faith or a Table of the most ancient creeds: engraved in Copper by Mr. Senex

Not seen. Not in B.M., C.U.L. or Clare College, Cambridge.

1713

Three Essays

These three essays were issued as one work in 1713. The second and third essays were also sold separately.

The titles are as follows:

I. The Council of Nicea vindicated from the Athanasian Heresy

II. A Collection of Ancient Monuments relating to the Trinity and Incarnation, and to the History of the Fourth Century of the Church

III. The Liturgy of the Church of England reduced nearer to the Primitive Standard

I. The Council of Nicea vindicated from the Athanasian heresy (1713)

This short work, 28 pages, sets out to show that the Primitive faith of the Christian Church concerning the doctrines of the Trinity and the Incarnation differs from the formulation of these doctrines as presented by Athanasius. Twenty three Articles of Primitive Faith concerning either the Trinity or Incarnation are presented. In each case the article is followed by the Athanasian doctrine of the same article.

Whiston's sympathy evidently lies with the doctrine held by that body of Christians later termed Eusebians who allowed the Divinity of the Son but were not happy to use words signifying the same essence and substance applied to the Father and the Son.

Whiston's final conclusion is to regret the imposition of the Athanasian creed as the profession of the Christian faith in the Trinity and the Incarnation.

II. **A Collection of Ancient Monuments relating to the Trinity and Incarnation, and to the History of the Fourth Century of the Church** (1713)

This work of 220 pages of very small print is a compilation of source material intended to give one a background of authentic documents against which one could assess the true contributions of Arius, Eusebius and Athanasius to the Council of Nicea. The work consists of forty-five separate items most of which are from the English translation of Eusebius, and Socrates' Historia Ecclesiae printed in English at Cambridge in 1683. This collection of sources was meant to be used in conjunction with the foregoing essay, The Council of Nicea vindicated from the Athanasian heresy where we read

> "... all of which assertions are fully proved
> by the authentic records of the same Council,
> most of which I have published in the ensuing
> treatise."

III. **The Liturgy of the Church of England reduced nearer to the Primitive Standard**

(already mentioned).

1714

An Argument to Prove that either all Persons solemnly though irregularly set apart for the Ministry are real Clergy-Men and all their Ministerial acts are valid; or else there are no real Clergy-Men or Christians in the World (1714)

Whiston argues that there are now no clergy who were appointed by our Saviour and his Apostles. The theme of the essay is a plea for toleration and acceptance of the goodwill of Christians of all denominations.

Two appendices to this work are

(1) To the Archdeacon Hill of Wells.

(2)   A Reply to Mr. Thirlby's Second Defence of Athanasius.

1715

A Vindication of the Sibylline Oracles to which are added
the genuine oracles themselves; with the ancient citations
from them; in their originals, and in English

The Judaeo-Christian Sibylline Oracles were regarded
by some scholars as a prophetic authority comparable to
the Old Testament.   Modern scholars have been able to
assign dates to the various oracles, largely by comparing
the known sequence of events with what the oracles predicted.   In 1713, Sir John Floyer published a translation
of the Sibylline Oracles from the best Greek copies and
compared them with Sacred prophecy especially in Daniel
and the Revelations.

Whiston's 1715 publication was dedicated to the notable
classical scholar Richard Bentley.   Whiston gives Floyer's
translation, 'but so much corrected by myself as to be in
a manner equivalent to a new translation.'   A Manuscript
note in the British Museum copy of this work (Shelf Number
873. 1.23) recognises the help received from his son-in-law,
Samuel Barker.   Whiston was doubtless influenced by the
fact that Flavius Josephus reports that the Jews took notice
of the oracles in undertaking their wars.   The Apostolical
Constitutions also quote the Oracles for the confirmation
of a number of Christian doctrines such as the resurrection
of the body and the existence of rewards and punishments
hereafter.   Whiston's interest in the Sibylline Oracles
is therefore seen to be connected with his studies in
chronology, sacred antiquity and primitive Christianity.

1716

St. Clement's and St. Irenaeus's Vindication of the Apostolical Constitutions (1716)

To this short essay are added Two Ancient Rules for the celebration of Easter also a Postscript prompted by Mr. Turner's discourse of the Apostolical Constitutions. A second edition of this work was printed in 1716 to which was added a Supplement with further notes and commentaries on the Apostolical Constitutions. An interesting note gives evidence of Whiston's continued interest in the literal fulfilment of Scripture prophecies. He claims that several events of the years 1715 and 1716 are the fulfilment of the coming to an end of "the wilderness condition" which Whiston interprets as the entirely obscure and persecuted state of the Church. Some of these events he names, the death of the King of France, Aug. 21 1716, the advancement of the Cardinal de Noailles; the greater freedom of the Reformed and the release of the Protestants in the Gallies.

The British Museum copy of this work (Shelf Number 873. 1.1(7)) contains many manuscript notes by the author.

1716

An Humble and Serious address to the Princes and States of Europe for the Admission or at least Open Toleration of the Christian Religion in their Dominions (1716)

This work is written from Yelvertoft in Northamptonshire. It sets out to show that the various European countries do in fact only support one particular sect of Christianity. Some parts of Germany only support the set of doctrines and rules agreed on by the Lutherans in 1530. He urges that any professed Christian state should support the practice and doctrine of early forms of Christianity which Whiston identifies with that found in the Apostolical Constitutions.

1717

Scripture Politicks: or an Impartial Account of the Origin and Measures of Government Ecclesiastical and Civil (1717)

This work was addressed to Benjamin Hoadley (1676-1761) then Bishop of Bangor and successively Bishop of Hereford, Salisbury and Winchester. Probably no cleric in the Church of England has been more violently attacked than Hoadley. As the prominent and aggressive leader of the extreme latitudinarian party in church and state, he naturally attracted the strongest assault of the Tory and high church party. As a minimiser of mysteries and dogma he was offensive to Churchmen. He incurred criticism for the practice of abuse of benefices. He never visited the diocese of Bangor and probably not that of Hereford.

The Bangorian Controversy started as the result of Hoadley's publication of 1716, A Preservative Against the Principles and Practices of the non-Jurors both in Church and State. On 31st March 1717, Hoadley preached a sermon before the King, on The Nature of the Kingdom or Church of Christ, denying that there is such a thing as the visible Church of Christ. Whiston's Scripture Politicks was one of nearly two hundred contributions to the controversy.

A postscript added to this work states Whiston's firm conviction that a legal supremacy in ecclesiastical affairs is no guarantee of supremacy in the spiritual laws of the Gospel. This sentiment agrees with that expressed by

Whiston in his address to the Bishop of Bangor in which he deplores the action of the State in its deprivation of the non-juring Bishops at the Revolution of 1688.

To this work, Whiston adds a small Supposal (8 pages) also entitled a New Scheme of Government which was first published in 1712.   Whiston advocates that Bishops and Clergy are to manage all the spiritual affairs of each diocese and no ecclesiastical persons will be permitted to participate in the strictly public affairs of Her Majesty's Government.

1718

A Primitive Catechism by way of Question and Answer in Two Parts useful for Charity Schools with the texts of Scripture proper for the proof of the several answers

The author of this work is Daniel Whiston, youngest brother of William Whiston under the title A Primitive Catechism.   In 1718, Whiston edited the work and published it under the title given above.   The work is divided into two parts.   Part I, for the Catechumens or Beginners in Christianity;   Part II for the Illuminates or Candidates for Baptism.   Whiston's publication of 1718 includes the catechism of Daniel Whiston which is followed by a summarised version in which the texts of scripture are omitted. This was probably meant for younger members of the charity schools.

A second, enlarged version of this catechism was published in 1751.   The British Museum copy of this work (Shelf No. 873, 1.21) includes two pages of manuscript by Whiston giving revised answers to the questions relating to

the celebration of the Eucharist, the dispositions of the
participants and the benefits to be expected from worthy
reception of the Eucharist.

1718

A Primitive Catechism by way of question and answer in two
parts useful for Charity Schools with texts of Scripture
(1718).

This work is discussed in Chapter V of the text.

1719

Mr. Whiston's Letter of Thanks to the Right Reverend the
Lord Bishop of London, for his late Letter to his clergy
against the Use of New Forms of Doxology (1719)

The Bishop of London to whom this letter was addressed
was John Robinson.   Robinson had been warning his Clergy
against new forms of Doxology agreeable to ancient heretics.
Whiston challenged him that, in fact, the doxologies in use
by the Church of England were in fact new.

This letter went through three editions in 1719.

In the same year Whiston published a further letter
entitled, A Second Letter to the Lord Bishop of London concerning the Primitive Doxologies (1719).

In this second letter Whiston answers the publication
of C. Paets which was occasioned by Whiston's first letter.

1719

Mr. Whiston's Account of Dr. Sacheverell's Proceedings in
order to exclude him from St. Andrew's Church in Holborn
(1719).

Henry Sacheverell (1674(?)-1724) was elected chaplain
of St. Saviour's Southwark in 1705.   In November 1709,
Sacheverell preached at St. Paul's before the Lord Mayor and

Aldermen on "the perils of false brethren in church and state". He declared the Church in danger from toleration and openly attacked the attitude of the Bishop of Salisbury, Benjamin Hoadley. The House of Commons ordered that Sacheverell should be impeached for malicious and scandalous libels which reflected on Her Majesty and her (Whig) Government. He was suspended from preaching for three years. The feeling of the country was, however, strongly on Sacheverell's side and forty thousand copies of the St. Paul's sermon were supposed to have been circulated.

Sacheverell resumed his preaching from 1713 and Queen Anne gave him a rich living of St. Andrew's, Holborn. In 1719, Whiston was one of his parishioners and Sacheverell made a public demonstration to have Whiston removed from his church because of his Arian beliefs.

In this publication, Whiston gives a graphic account of several such incidents.

1719

Mr. Whiston's Letter to the Right Honourable the Earl of Nottingham, concerning the Eternity of the Son of God and of the Holy Spirit (1719)

This letter is addressed to Daniel Finch, second Earl of Nottingham, because he had recently been active in opposing a clause for the Toleration of the Christian Religion itself or of all that believed the Holy Scriptures and the Common Creed, in a debate in the House of Peers. The Earl of Nottingham had strongly supported Henry Sacheverell in his declaration that the Church was in danger from toleration, occasional conformity and schism. Finch had evidently

censured persons for omitting the Holy Spirit in Doxologies.
Whiston proceeds to examine the doctrine of the Trinity and
to show that the co-eternity of the Son and Holy Spirit do
not appear to be the doctrine of the early Church.   Whiston
quotes largely from Vol. IV of <u>Primitive Christianity</u>
<u>Revived</u>.   Article vi of this volume is entitled, <u>The Texts</u>
<u>and Testimonies for the original voluntary Generation and</u>
<u>Creation of the Son of God before the World began;   and</u>
<u>against his Co-Eternity with the Father</u>.

This letter appeared as a second edition and third
edition in 1721 with a large Preface (76 pages) on the
occasion of the Earl of Nottingham's Answer to that letter.
This Preface makes full and detailed references to the
literature of the seventeenth and early eighteenth centuries,
concerning the doctrine of the Trinity.   Whiston owns that
his own education was based on commentators such as Bishop
Pearson, Bishop Simon Patrick and Bishop Kidder.   Whiston
is certainly familiar with the controversialists who uphold
the Athanasian formulation, Bishop Bull, Dr. Grabe, Dr.
Daniel Waterland and Du Pin.   Here as elsewhere, Whiston
leans heavily on the fathers of the early church, Origen,
Eusebius, Irenaeus, Athenagoras.

In addition to the 1721 editions is <u>Athanasian</u>
<u>Confessions</u>, that the Antenicene writers were against
the Athanasian and for the Eusebian doctrine.

1719

A Commentary on the Three Catholic Epistles of St. John in agreement with the Ancient Records of Christianity now Extant (1719)

This work sets out to prove that the three Catholic Epistles of St. John are authentic and agree with the most ancient records of Christianity.   An Appendix in the form of thirteen Articles forms a collection of ancient testimonies, useful for the understanding of these epistles.

Modern scholars admit the Apostolic authorship of the first Epistle which opposes the false doctrines on the person of Christ.   The other two epistles are intimately connected and are not generally admitted as authentic.

1720

The True Origin of the Sabellian and Athanasian Doctrines of the Trinity or a Demonstration that they were first broached by the Followers of Simon Magus, in the First Century and Revived by the Montanists in the second (1720)

This work was recommended by Whiston to the consideration of Rev. Daniel Waterland.   Waterland was a staunch defender of the orthodox doctrine of the Trinity and had written a masterly defence of the doctrine in reply to the writings of Samuel Clarke.   Waterland was also the author of a History of the Athanasian Creed.

Whiston's arguments against the Athanasian doctrine of the Trinity are based largely on the writings of Tertullian among other ancient testimonies.   Whiston also tries to elucidate the origin of Sabellian doctrine of the Trinity.   Sabellius acknowledged the Father, Son and Holy Spirit to be the same Person.

1721

Mr. Whiston's Reply to the Right Honourable the Earl of Nottingham (1721)

This title which appears in catalogues of Whiston's work is in fact a title page to the second and third editions of Whiston's, Letter to the Right Honourable the Earl of Nottingham, concerning the Eternity of the Son of God and of the Holy Spirit already discussed.

Whiston's reply to the Earl of Nottingham is contained in the extensive Preface to later editions of the 1719 letter.

1721

A Chronological Table of the Hebrew, Phoenician and Chaldean Antiquities, compared together. This chart belongs as an Appendix to An Essay towards restoring of the True Text of the Old Testament (1722).

Not seen. Not in B.M., C.U.L., Clare College, Cambridge.

1722

An Essay towards restoring the True Text of the Old Testament (1722)

This is a considerable work of 333 pages laid out under thirteen propositions. It aims to establish changes which have emerged through successive different versions of parts of the Old Testament. Some of the later propositions aim to correlate Old Testament events with secular history and in particular with the works of the Jewish historian Josephus.

A large Appendix to this work falls under four headings:

1.   The Variations of the Samaritan Pentateuch from the Hebrew.

2.   A Demonstration that the Apostolical Constitutions were written in the First Century.

3.   That Sesostris was that Pharaoh who was drowned in the Red Sea.

4.   A Collection of Original Monuments referred to in my Chronological Table.

Whiston published a Supplement to this work in 1723 entitled, A Supplement to Mr. Whiston's late Essay, towards restoring the True Text of the Old Testament proving that the Canticles is not a sacred book of the Old Testament; nor was originally esteemed as such, either by the Jewish or Christian Church (1723).

This small work (50 pages) is a plea that the Canticles of Solomon should be disregarded as a sacred book. The two main arguments for this view are firstly Solomon's pre-occupation with his thousand women, 700 Queens and 800 Concubines during the period when he wrote this work and secondly the fact that the Canticles is not mentioned in the Apostolical Constitutions as a sacred book.

The British Museum copy of this work (Shelf Number 873, 1.10) contains many manuscript notes by the author.

1724-1725

The Literal Accomplishment of Scripture Prophecies (1724)

This work was written in answer to the publication of Anthony Collins (1676-1729), A Discourse on the Grounds and Reasons of the Christian Religion, (1724). Collins was a close friend of John Locke during the last few years of Locke's life. He published several controversial works

which urged that all belief should be based upon free
enquiry.   Collins' 1724 work is a direct attack on Whiston's
allegorical interpretation of scripture prophecy.   Collins
was one of the most conspicuous of the Deist writers who
took the line of historical criticism.

The first five sections of this work instance further
literal fulfilment of Scripture prophecies.   Whiston links
the fulfilment of the third Sibylline Oracle,

> "All the paths of the field, and rough shores,
> and high mountains, and the raging waves of the
> sea shall be easily passed over and sailed over
> in those days,"

with the discovery of the mariner's compass, the knowledge
of its variation, the variation of its variation, the
phenomenon of the dipping needle.   By all these means,
travelling and sailing are easier and more practicable than
formerly.

An Appendix to this work is entitled, <u>Aristeas's
History of the Version of the Pentateuch by the LXXII
interpreters, still extant, is genuine</u>.

Also included in this work is a list of Suppositions
or assertions in the late <u>Discourse of the Grounds and
Reasons of the Christian Religion, which are not therein
supported by any real or authentic evidence; for which
some such evidence is expected to be produced</u>.

A Supplement to the <u>Literal Accomplishment of
Scripture Prophecies</u> published in 1725 opens with observations on Dr. Samuel Clarke's and Bishop Chandler's discourses of the prophecies of the Old Testament.   Whiston
is not able to accept the view put forward by Clarke and

Chandler that some scripture prophecies are to be understood in a double sense. The basis of all Whiston's writings with regard to prophecy is that they have a single meaning and will in due course be literally fulfilled.

The final section of the supplement is divided into four dissertations:

1. Upon Isaiah's Prophecy concerning a Son to be born of a Virgin.

2. Upon Daniel's LXX Weeks.

3. Upon the IVth Eclogue of Virgil as compared with the Sibylline Oracles.

4. Upon the Curses denounced against Cain and Lamech before the Flood; proving that Africans and Indians are their posterity.

The British Museum copy of this work (Shelf Number 873. 1.16(1)) 873. 1.16(2)) contains manuscript notes and corrections by the author.

1726

Of the Thundering Legion: or of the miraculous deliverance of Marcus Antoninus and his army upon the Prayers of the Christians (1726)

This short publication was prompted by a posthumous publication of the historian Walter Moyle (1672-1721). Moyle's works were published in two volumes in 1726 and included, Remarks on the Thundering Legion.

Another short work which forms part of Whiston's 1726 publication was Of Alexander the Great's meeting with the High-Priest of the Jews at Jerusalem.

A Collection of Authentic Records belonging to the Old and
New Testament (1727, 1728) in two volumes

In 1724 Whiston circulated proposals for Printing by
subscription Authentic Records concerning the Jewish and
Christian Religions, in three volumes 8vo.

A note at the end of these proposals summarises
Whiston's motive in publishing this work.   It is to counter
the influence of,

> "our present unbelievers who openly deny the truth
> of the Jewish and Christian revelations and the
> veracity of their sacred records ... as if they
> could disprove them by real and original evidence
> or by the genuine authentic testimonies of
> Antiquity."

This work finally appeared in 1727 and 1728 in 2
volumes under the slightly amended title given above, and is
a defence of the authenticity of the Jewish and Christian
records.

In his memoirs, Whiston states that until the publication of his Josephus (1737), he regarded this work as his
Opus Palmarium

Attached to the 1749 edition of his Memoirs, Whiston
published A scheme of the Seven Heavens corresponding to
the Courts of the Temple with its several parts.   This was
meant as an illustration of the Testament of Levi, one of
the twelve Patriarchs published in this work of 1727.

This publication of 1727 is of considerable size, 1123
pages, and is subdivided into 47 parts.   The table of
contents is listed in his Memoirs.   Appendix IX which forms
Section XXIV of volume II is A Confutation of Sir Isaac
Newton's Chronology.

1727

Mr. Henley's Letters and Advertisements which Concern Mr. Whiston

John Henley (1692-1756) was an M.A. of St. John's College, Cambridge, who later became an eccentric London preacher.   In 1726, he left the Church and rented rooms in Newport Market where he delivered sermons and orations. His publications, The Primitive Liturgy for the Use of the Oratory (1727), Letters and Advertisements (1727), and The Appeal of the Oratory and the First Ages of Christianity angered Whiston.

Henley forbade Whiston to enter his rooms at Newport Market.   Whiston accused Henley of giving scandal by the nature of his relationship with Mrs. Colson (or Tolson). There was an exchange of angry letters between Whiston and Henley which were printed in this pamphlet.

1729

Apostolical Rules for Ecclesiastical Courts (1729)

This small work, 24 pages, consists of the English translation of Book II, Chapter XXI of the Apostolical Constitutions which deals with judicial procedures.   It is followed by fourteen observations which indicate the divergence in practice of current judicial proceedings from the principles advocated in the Apostolical Constitutions.

The work is evidently inspired by current events.

Observation No. XIV notes,

"a foreign Heidegger is permitted to flourish at Court, and a foreign Facio is forced to stand in the pillory; whereby a native, John Henley escapes uncontrolled and a native Robert Hales, is convicted of forgery."

U.L.C. F.5.103.

1730

The Horeb Covenant Revived (1730)

The secondary title of this work, An Account of Those Laws of Moses which oblige Christians, explains its contents.

The main part of the work is printed in parallel columns headed, Laws of Moses and Parallel Laws of Christ. Whiston sets out to show that in many instances, Christ came to fulfil and complete the Law, not to destroy it. Whiston refers to similar sentiments expressed by Newton in his Chronology.

U.L.C. Ddd. 25.220.

1730

Historical Memoirs of the Life of Dr. Samuel Clarke (1730)

This work was occasioned by the death of Clarke in 1729.

It is not a life of Clarke but includes Whiston's account of his first and some subsequent meetings with Clarke. It expounds Clarke's views on the Trinity but is largely devoted to Whiston's correspondence about the source of his own views of the Trinity.

At the end of this work are listed twenty-two printed works of Clarke, also publications of Samuel Clarke from his manuscript by his brother John, Dean of Sarum.

This work went through several editions. The third edition, 1748, has an Appendix divided into two parts:

1. Dr. Syke's Elogium of Dr. Clarke.

2. Mr. Emlyn's Memoirs of the Life and Sentiments of Dr. Clarke.

1731

A Description of the Tabernacle of Moses and of the Temples of Jerusalem; in a very large open sheet (1731)

This chart, about 24" square, consists of five insets.

1. A scale drawing of Solomon's Temple.

2. The Tabernacle of Moses, Titus's Triumphal Arc with separate enlarged drawings of the Candlestick and Table of Shew Bread.

3. The Jewish High Priest with the Ark of the Covenant and the Altars of Burnt-offering and of incense.

4. The Perspective View or Elevation of Ezechiel's Court of the Priests and of the Holy House itself, with separate drawings of the Brazen Sea, the curious chapiters of Jachin and Boaz and a Jewish shekel.

5. Mr. Sandy's Map of Jerusalem.

This chart was also printed in the 1737 edition of the Josephus.

1732

The Testimony of Phlegon vindicated: or an account of the great Darkness and Earthquake at our Saviour's Passion described by Phlegon (1732).

In this short work, Whiston sets out to show that Phlegon's famous testimony concerning a great eclipse and earthquake was in fact the event referred to by the four evangelists. Phlegon was a native of Tralles in Lydia, a freedman of the emperor Hadrian, one of whose works was a treatise on wonderful events (including the eclipse and earthquake mentioned). Whiston calls on the original testimonies of authors of the first six centuries to substantiate the claim. This work is followed by Observations from the foregoing testimonies which Whiston makes to prove his claim.

1734

Six Dissertations (1734)

These are listed in the title page as follows:

1. The Testimonies of Josephus concerning Jesus Christ, John the Baptist, and James the Just vindicated.

2. The Copy of the Old Testament made use of by Josephus proved to be that which was collected by Nehemiah.

3. A Reply to Dr. Syke's Defence of his Dissertation on the Eclipse mentioned by Phlegon. (In the text entitled Appendix I).

4. The Chronology of the Sacred Scriptures and the Truth of their Predictions confirmed by Eclipses and Astronomical Observations. (In the text entitled Appendix II).

5. Remarks on Sir Isaac Newton's Observations upon the Prophecies of Daniel and the Apocalypse. (In the text entitled Appendix III and subdivided into Parts 1 and 2).

6. A Demonstration that our Saviour's Ministry continued at least Four Years, occasioned by a late Dissertation on that subject. (In the text entitled Postscript. 9 pages).

Dissertation 4 is a remarkably detailed account (88 pages) applying the science of astronomy to verify Biblical chronology. Whiston refers to the astronomical works of Halley and Cassini and to attempts to determine the antiquity of the Chinese Annals from Astronomy. Whiston also refers to Halley's attempts to extend the age of the earth beyond that indicated in the Biblical account by considering the increased saltness of the sea. (Philosophical Transactions, 1715).

1734

A Dissertation concerning God's command to Abraham to offer
up Isaac his Son as a Sacrifice (1734)

Whiston links the time of abolition of human sacrifice in Egypt with prohibition of by God of Abraham's sacrifice of his son Isaac.

1736

An Enquiry into the Evidence of Archbishop Cranmer's
Recantation: or reasons for a suspicion that the pretended
copy of it is not genuine (1736)

This short work (21 pages) was written in Lyndon,
Rutland in 1732. It is followed by a postscript written by
Whiston in 1736 in which he claims that this enquiry was
written by him before the publication of Mr. Strype's
Memorials of Archbishop Cranmer. There follows an
Advertisement (12 pages) written by J.A. on Mr. Strype's
account of Archbishop Cranmer's Death.

1736

Athanasian Forgeries, Impositions and Interpolations (1736)

This work was written under the pseudonym of
A Lover of Truth and True Religion. In his Memoirs, Whiston
acknowledges that he himself was the author of this pamphlet
of 115 pages. The work is organised under twenty-two
headings and consists largely of extracts from Whiston's
own earlier writings on the doctrine of the Trinity in the
early church and the part played by Athanasius the Council
of Nicea.

1736

The Primitive Eucharist Revived or an Account of the Doctrine
and Practice of the Two First Centuries concerning the
celebration of the Lord's Supper (1736)

Whiston sees this work as Part Two of the small
treatise Primitive Infant Baptism Revived (1712). The
title page of the 1736 work states that the publication is
occasioned by a recently published treatise, A Plain Account
of the Nature and End of the Sacrament of the Lord's Supper.
Whiston devotes 76 pages to a study of the original Texts
and Testimonies which tell of the doctrine and practice of
the two first centuries. A series of twenty-six observations follow in which Whiston defends the teaching of the
early Church and discusses this in the light of current
writings on the Eucharist and referring to such works as
Mr. King's Constitution, Discipline, Unity and Worship of
the Three First Centuries and those of Dr. Lightfoot,
Bishop Hooper, Dr. Grabe and Mr. Mede.

1737

The Genuine Works of Flavius Josephus the Jewish Historian
(1737)

The title page of this work is reproduced in full as
it gives details of all Whiston's writings on Josephus
gathered together in two large folio volumes (1021 pages).
The Genuine Works of Flavius Josephus the Jewish Historian
translated from the original Greek according to Havercamp's
accurate edition, containing twenty books of the Jewish
Antiquities, with the Appendix or Life of Josephus written
by himself: Seven Books of the Jewish War and Two Books
against Apion.

Illustrated with new plans and descriptions of the Tabernacle of Moses; and of the Temples of Solomon, Zorobabel, Herod and Ezekiel; and with correct maps of Judea and Jerusalem.

Together with

Proper notes, Observations, Contents, Parallel Texts of Scripture, five complete indexes, and the true chronology of the several Histories adjusted in the Margin.

To this book are prefixed eight Dissertations viz.

1.   The Testimonies of Josephus vindicated.

2.   The copy of the Old Testament made us of by Josephus proved to be that which was collected by Nehemiah.

3.   Concerning God's Command to Abraham to offer up Isaac his Son for a Sacrifice.

4.   A large enquiry into the true Chronology of Josephus.

5.   An extract out of Josephus's Exhortation to the Greeks, concerning Hades, and the Resurrection of the Dead.

6.   Proofs that this Exhortation is genuine; and was no other than a Homily of Josephus's, when he was Bishop of Jerusalem.

7.   A Demonstration that Tacitus, the Roman historian, took his history of the Jews out of Josephus.

8.   A Dissertation of Cellarius's against Harduin; in vindication of Josephus's History of the family of Herod from Coins.   Translated into English.

With an account of the Jewish Coins, Weights and Measures.

Whiston's Josephus was printed by W. Bowyer and was sold by John Whiston, Bookseller at Mr. Boyle's Head: Fleetstreet.

The Map of Palestine is attached after the title page. It measures 22" x 24" approx. and is entitled Cellarius's, Peland's and Maudrell's Maps of Palestine corrected and improved by Mr. Whiston, with the addition of an alphabetical index of the several places with their longitudes East or West from Jerusalem and their latitudes north: for the ready finding of them in the Map.

Whiston explains that he follows Mr. Morden and Dr. Halley in placing Jerusalem 32° of North Latitude and Dr. Halley in placing it 2 hours 23 mins. or $35\frac{3}{4}$° Longitude from Greenwich, England.

The first two of the eight dissertations published as part of this work had already been printed as the first two of the Six Dissertations published by Whiston in 1734.

Whiston was justly proud of his monumental work on Josephus and at the end adds this note.

N.B.   I began this version (after I had frequently
         perused Josephus in the original, and prepared the
         preliminary Dissertations and the Notes beforehand)
         on December 9th A.D. 1734, the day that I was 67
         years of age: and finished it on January 6th 1737
         in the beginning of my 70th year; having been two
         years and one month about it.

The illustration of the Tabernacle of Moses and the temples of Jerusalem etc. which was included in this work is identical with the publication already described in 1731.

1737

An Account of the Demoniacs and of the Power of casting out Demons both in the New Testament and in the Four First Centuries (1736)

This work was prompted by a recent publication, An Enquiry into the meaning of Demoniacs in the New Testament. In this work Whiston defends what he considers to be the tradition of the Church regarding the existence of a personal devil, Satan.

To this work is added an Appendix,

A Collection of Original Texts of Scripture, and Testimonies
of Christian Antiquity, concerning Tythes and Oblations,
for the Maintenance of the Poor, and the Clergy in the first
four centuries.

1740

The Eternity of Hell Torments Considered or a collection of
texts of scripture and testimonies of the three first
centuries relating to them together with notes through the
whole and observations at the End

Whiston acknowledges that his doubts concerning the
eternity of hell torments were expressed as early as 1709
in his Sermons and Essays under the title Reason and
Philosophy no Enemies to Faith. Whiston also assures the
reader that he is not going to undertake a study about the
places of Hades and Gehenna. This problem is now solved
due to the recent discoveries of Halley and Whiston concerning Terrestrial Magnetism. The cavity between the
loadstone at the core of the earth and the outer sphere
naturally corresponds to the places of Hades and Gehenna!

Whiston sets out to prove that the teaching of the
Old and New Testament, and of the Church up to the Council
of Nicea, never entertained the notion of eternal punishment
of the wicked.

1741

Mr. Whiston's Appeal to the Thirty Primitive Councils Against
the Athanasian Heresy (1741)

This small pamphlet was written as an Appendix to
Athanasian Forgeries, Impositions and Interpolations (1736).
Whiston's survey of the proceedings of the Councils is based
on Cave's Historia Literaria and Du Pin's Bibliotheca.

Whiston traces deliberations relating to the Doctrine of the Trinity between the years 260 and 360 A.D.

<u>1745</u>

<u>Mr. Whiston's Sacred History of the Old and New Testament</u>
from the creation of the World till the days of Constantine the Great (1745) (3 vols.)

In the Preface of this work Whiston acknowledges that six parts out of seven are taken from the work of his old friend Humphrey Prideaux, <u>The Old and New Testament connected in the History of the Jews and the neighbouring nations</u> (2 vols. 1716-1718).

Whiston presumes to put his own name to this work because he judges that his corrections and improvements are considerable and greatly add to the completion of the history. Whiston has also consulted Samuel Shuckford's work, <u>The Sacred and Profane History of the World, Connected</u> (1728), and also Rollin's valuable accounts of ancient times.

Whiston considered it necessary to improve on the works already mentioned since none of them started their account with the Creation. Prideaux's work adopted Ussher's chronology to which Whiston has made amendments according to his own discoveries. Whiston feels able to improve certain aspects of Eusebius's <u>Ecclesiastical History</u> which deals with the three first centuries of the Church as a result of his studies on Josephus, the Jewish historian.

Volume 2 closes with a detailed <u>Chronological Index from the Creation till the Birth of Christ</u>. Volume 3 ends with a <u>Chronological Index of Ecclesiastical Affairs from the Nativity of our Saviour Jesus Christ till the days of Constantine the Great</u>.

Pages 609-613 of Volume 3 give details of Proposals
issued by Whiston Feb. 26, 1738, for printing a cheap edition of the Primitive Fathers up to the middle of the Fourth.
Century and the distributing of these to parochial libraries,

"In every parish of Great Britain, Ireland;
with the Plantations thereto belonging."

Whiston esteemed this work,

"as the last great work of my life ... because
it is to take in all the remarkable histories
that directly concern the Jews ... It is an
entire abridgement of and supplement to the
Sacred Books of the Bible, to Josephus's
Antiquities, to the Synopsis Sacrae Scripturae
and to Eusebius's Chronicon, Ecclesiastical
History and Evangelical Preparation." (Preface)

1745

Mr. Whiston's Primitive New Testament (1745)

This is divided into Four Parts.

1. Containing the Four Gospels with the Acts of the
Apostles.

2. Containing 14 Epistles of Paul.

3. Containing 7 Catholic Epistles.

4. Containing the Revelation of John.

In each part, Whiston gives details of the manuscript
or other source from which the translation was made.

Bound to the copy consulted (Clare College, Cambridge)
was a short essay on the Resurrection of Jesus Christ,
according to Beza's double copy of the four Gospels and
Acts of the Apostles.

It seems likely that Whiston intended to publish a
companion volume to this work consisting of four further
parts in which are listed the rest of the books of the New
Testament which, as he says in his memoirs, p.386, were not

as yet known by most Christians.   This list as would be expected included the <u>Constitutions of the Apostles</u>.

1748

<u>Mr. Whiston's Friendly Address to the Baptists</u> (1748)

This publication was written at Lyndon, Rutland in 1747 and appears in Part II of his Memoirs published in 1749. Whiston examines his reasons for being so strongly attracted to the Baptists.   He acknowledges that he has been driven from the Church of England by its use of the Athanasian creed.   Since Trinity Sunday 1747, Whiston went out of the Church of Lyndon never to return to it.

Whiston chooses the Baptists as nearest to the primitive practice of Christ and the Apostles.   Whiston then proceeds to compare the original baptismal creed found in the <u>Apostolical Constitutions</u> with the creeds of Athanasius and Pope Pius V (?).   This address was also printed in 1748 and sold separately.

1749

<u>Mr. Whiston's Account of the Exact Time when Miraculous Gifts ceased in the Church</u> (1749)

This short publication was first printed in 1728 as part of the work <u>A Collection of Authentic Records belonging to the Old and New Testaments</u>.   Whiston believed, on the authority of Chrysostom, that miraculous gifts ceased about 384 or after the Council of Constantinople.

It was printed again in 1749 in Part 2 of Whiston's Memoirs with Observations upon Dr. Middleton's Free Inquiry into that Matter.

Memoirs of the Life and Writings of Mr. William Whiston containing Memoirs of several of his Friends also (1749)

Whiston's Memoirs are a rich source of information concerning his religious ideas and publications. Many of the letters, sermons and prayers which are found in the Memoirs are reproduced among his printed works. A second edition of the Memoirs appeared in 1753, the year after Whiston's death. The Memoirs appear as three separate volumes or as two volumes, the first of which includes Part 1 and 2 of the Memoirs.

BIBLIOGRAPHY

1. MANUSCRIPTS

Listed below are the Archives and Record Repositories holding manuscripts relating to William Whiston which the author has consulted in the preparation of this Thesis. Detailed description of each manuscript item has been avoided.

Beinecke Library, Yale University
Letters relating to Whiston's scientific works.

Bodleian Library, Oxford
A varied collection of manuscripts including Whiston's Commonplace book and the Minutes of the Society for Promoting Primitive Christianity.

British Museum
Add. MS. 24197.   Volume of Correspondence.
Whiston and William Lloyd.
Burney MS. 522. F.2.
Notes of Astronomical Lecture.
Add. MS. 45511 and 28104
Letters.

Cambridge University Library
Letter - Whiston to Strype.

Clare College, Cambridge
Manuscript version of Whiston's 'mock-trial'.
Fellows Accounts 1684-1718 including Whiston's entries.
Several letters in the College letter book.

Dr. Williams' Library, 14 Gordon Square, London W.C. 1.
Baxter's letters, Vol. VI.
Josiah Whiston to Richard Baxter.

Gloucestershire Record Office
D 227/19.
Notes by Annie Raine Ellis among the Ellis and Viner family papers.
A.R. Ellis made copious notes in preparation for a biography of Whiston.   Her only source was Whiston's Memoirs.

Leicester Museums Department of Archives
Bishops Transcripts of parish register for Norton-juxta-Twycross.

Leicestershire Record Office

Among the manuscripts of the Reverend Edmund Field is the Barker Correspondence. This latter includes the Whiston Papers. This interesting collection is so far unpublished and the author has made extensive use of some 70 items. Manuscripts relating to Whiston's work on the Longitude form part of this collection. These papers are privately owned by Mr. J. Conant of Lyndon Hall, Rutland.

Legal Documents
1018. Will of William Whiston.
1021. Will of Samuel Barker.
1024. Will of George Whiston.
      Will and Inventory of Josiah Whiston.

National Library of Scotland

Letter (1747) relating to Whiston and Hoadly.

Old Royal Observatory Archives, Herstmonceux

Copies of all these records are kept in Flamsteed House, Greenwich Park.
Many references to Whiston's work on Longitude among the Flamsteed Papers.
Entries relating to Whiston in the Minutes of the Board of Longitude.

Old Schools, Cambridge, University Archives

Documents relating to Whiston's banishment from Cambridge.

Parish Registers at:

Norton-juxta-Twycross in Leicestershire,
Drayton, near Norwich,
St. Margaret's, Lowestoft, Norfolk,
Lyndon, Oakham, Rutland.

Royal Commission on Historical Manuscripts

Occasional letters relating to Whiston among printed volumes of: Bath MS, Egmont MS, Portland MS.

Royal Society

Many entries in Journal Books, Record Books and Letter Books relating to Whiston's scientific work.
Classified Papers Vol. XX
Abstract of Whiston's New Theory of the Earth
Fo. 4.28.
Letter Whiston to Folkes.

Trinity College, Cambridge

Dawson Turner Collection.
Letters relating to Whiston's work on Longitude.

University Library of North Wales

Manuscript relating to Whiston's controversy with the Bishop of Bangor.

2.   PRIMARY SOURCES

The printed works of William Whiston.

Maps and charts of William Whiston.
(Map Room, British Museum).

Allix, Peter.
     Remarks upon some places in Mr. Whiston's Books
     second edition (1711).

Bentley, Richard.
     Boyle Lectures for the Confutation of Atheism (1693)

Burnet, Thomas.
     The Sacred Theory of the Earth, 6th edition (1726).

Burton, W.
     Description of Leicestershire (1622).

Cave, W.
     Primitive Christianity, seventh edition (1728).

Cotes, R.
     Hydrostatical and Pneumatical Lectures, second edition,
     (1747).

Descartes, Renee.
     Principia Philosophiae (1644).

Ditton, H.
     The General Laws of Nature and Motion (1705).

Dugdale, W.
     Short View of the Late Troubles in England.(1681).

Gellibrand, H.
     A Discourse Mathematical on the Variation of the
     Magnetic Needle (1635).

Gilbert, William.
     De Magnete (1600), Dover Publications (1958).

Hauksbee, F.
     Physico-Mechanical Experiments (1709).

Hawkins, Isaac.
     Essay for the Discovery of the Longitude at Sea (1714).

Hevelius, J.
     Epistolae, Gedani (1654).

Hooke, Robert.
     Discourse on Earthquakes (posthumous) (1705).

Keill, John.
An Examination of Dr. Burnet's Theory of the Earth with Remarks on Mr. Whiston's New Theory (1698).

Introductio ad Veram Physicam (1701).

Introductio ad Veram Astronomiam (1718).

Lovell, Archibald.
A Summary of Heads which may be enlarged into a Complete Answer to Dr. Burnet's Theory of the Earth (1696).

Newton, I.
Mathematical Principles of Natural Philosophy. Translated by A. Motte (1729), C.U.P. (1934).

Optics (1704).

Pepys, S.
Diary, edited H.B. Wheatley, London (1952).

Plot, Robert.
Natural History of Oxfordshire (1677).

Prideaux, H.
The Old and New Testament connected in the History of the Jews and Neighbouring Nations (1718).

Ray, John.
The Wisdom of God Manifested in the Works of Creation (1691).

Three Physico-Theological Discourses (1693).

Robinson, Thomas.
New Observations on the Natural History of the World (1696).

Saint Clair, Robert.
The Abyssinian Philosophy Confuted (1697).

Steno, Nicholas.
Prodromus (1669).

Stillingfleet, Edward.
Origines Sacrae (1662).

Warren, Erasmus.
Geologia (1690).

Waterland, D.
Advice to a Young Student, third edition (1760).

Woodward, John.
Essay towards a Natural History of the Earth (1695).

Worster, B.
A Compendious and Methodical Account of the Principles
of Natural Philosophy (1722).

3. SECONDARY SOURCES

Berry, A.
A Short History of Astronomy, Dover (1961).

Bosher, R.S.
The Making of the Restoration Settlement (1951).

Clark, G.
The Later Stuarts (1660-1714), second edition (1965).

Costello, W.T.
The Scholastic Curriculum at Seventeenth Century
Cambridge (1958).

Cohen, I.B.
Introduction to Astronomical Lectures by William
Whiston, Johnson Reprint Co. (1972).

Introduction to Sir Isaac Newton's Mathematick
Philosophy more Easily Demonstrated by William
Whiston, Johnson Reprint Co. (1972).

Cohn, N.
The Pursuit of the Millennium, Paladin (1970).

Collier, K.C.
Cosmogonies of Our Fathers (1934).

Cragg, G.R.
From Puritanism to the Age of Reason.

Reason and Authority in the Eighteenth Century,
C.U.P. (1964).

The Church and the Age of Reason (1970).

Davies, G.L.
The Earth in Decay (1968).

Douglas, D.C.
English Scholars (1939).

Edwards, W.N.
The Early History of Palaeontology, B.M. (Natural
History) (1967).

Haber, F.C.
The Age of the World.   Moses to Darwin (1959).

Gould, R.T.
John Harrison and his Timekeepers.   National Maritime Museum (1959).

The Marine Chronometer (1923).

Harrison, W.J.
Life in Clare Hall Cambridge 1658-1713. (1958).

Heath, T.L.
Euclid's Elements, Dover (1956).

Hellman, G.
Neudrücke von Schriften und Karten (1895).

Hill, C.
The Century of Revolution, Sphere edition (1969).

Hodgkiss, A.G.
Discovering Antique Maps, Shire Publications (1971).

Holmes, G.
The Trial of Dr. Sacheverell (1973).

Johnson, F.R.
Astronomical Thought in Renaissance England (1937).

Jacob, M. Candee,
The Church and The Boyle Lectures.   Ph.D. Thesis, Cornell University (1969).

Koyré, A.
From the Closed World to the Infinite Universe.
Johns Hopkins Paperback (1968).

Kubrin, D.
Providence and the Mechanical Philosophy.   Ph.D. Thesis, Cornell University (1968).

Leach, A.F.
Educational Charters and Documents 1598-1909 (1911).

Matthews, A.G.
Calamy Revised (1933).

May, Commander W.E.
Alexander Neckham and the Pivoted Compass Needle.
Journal of the Institute of Navigation, Vol.VIII, No.3.

Mayor, J.E.B.
Cambridge under Queen Anne (1911).

Merton, R.K.
Science, Technology and Society in Seventeenth Century England. (1936).

More, L.T.
Isaac Newton, Dover (1962).

Namier, L.B.
The Structure of Politics at the Accession of George III (1929).

Nicolson, M. and Rousseau, G.S.
This Long Disease, My Life, Princeton (1968).

Plumb, J.H.
The Growth of Political Stability in England 1675-1725. Peregrine (1969).

Schofield, R.E.
Mechanism and Materialism. British Natural Philosophy in an Age of Reason (1970).

Scott, J.F.
The Scientific Work of René Descartes (1596-1650). Taylor & Francis (1952).

Sollberger, Edmond.
The Babylonian Legend of the Flood. British Museum. third edition (1971).

Tawney, R.H.
Religion and the Rise of Capitalism, Pelican (1972).

Taylor, E.G.R.
The Mathematical Practitioners of Hanoverian England (1714-1840) (1966).

The Mathematical Practitioners of Tudor and Stuart England, C.U.P. (1967).

Walker, D.P.
The Decline of Hell. Routledge, Kegan Paul (1963).

Wardale, J.R.
Clare College. Letters and Documents (1903).

Waters, D.W.
Galileo and Longitude, Pjysis, Anno VI (1964), Fasc. 3.

The Art of Navigation in England in Elizabethan and early Stuart Times (1958)

White, A.D.
A History of the Warfare of Science with Theology in Christendom (1896).

Whiteside, D.T.
   The Mathematical Papers of Isaac Newton, Volume 1
   (1967).

Willey, B.
   The Seventeenth Century Background, Pelican (1972).

   The Eighteenth Century Background, Pelican (1972).

Articles from Journals

Biswas, A.K.
   The Hydrologic Cycle.
   Civil Engineering. ASCE. April 1965, p.70-74.

Davies, G.L.
   Early British Geomorphology, 1578-1705.
   Geographical Journal, 132, (2), p.252-262, 1966.

   Robert Hooke and his Conception of Earth History.
   Proceedings of the Geologists Association, 1964,
   Vol.75, part 4, p.493-498.

   The Concept of Denudation in Seventeenth Century England.
   Journal of the History of Ideas, 1966, Vol.27, No.2,
   p.278-284.

Eyles, V.A.
   John Woodward F.R.S. (1665-1728) Physician and Geologist
   Nature, 1965, Vol.206, p.868-870.

Gilmour, J.S.L.
   Huntia, 2 (1965).

Mitchell, A.C.
   Chapters in the History of Terrestrial Magnetism.
   Terrestrial Magnetism and Atmospheric Electricity,
   Vol.42, No.3, p.244.

Raven, C.E.
   Organic Design: A Study of Scientific Thought from
   Ray to Paley.
   Friends of Dr. Williams' Library, Seventh Lecture 1953.

Stokes, Evelyn.
   The Six Days and the Deluge.
   Earth Science Journal, Vol.3, No.1, 1969, p.13-39.

Taylor, E.G.R.
   The English Worldmakers of the Seventeenth Century and
   their Influence on the Earth Sciences.
   Geographical Review, 1948, 38, p.104-112.

Taylor, E.G.R.
   The Origins of Continents and Oceans: A Seventeenth
   Century Controversy.
   Geographical Journal, 116, 1950 (4-6), p.193-198.

Thompson, S.P.
   Petrus Peregrinus de Maricourt and his Epistola de
   Magnete.
   Proceedings of the British Academy, Vol.II, 1906.

Urey, H.
   Nature, March 2nd, 1973.

Waters, D.W.
   Time, Ships and Civilisation.
   Antiquarian Horological Journal, June 1963.

White, G.W.
   John Keill's View of the Hydrologic Cycle, 1698.
   Water Resources Research, December 1968, Vol.4, No.6,
   p.1371-1374.

4.   WORKS OF REFERENCE

Biographia Britannica (1766).

Biographical and Historical Dictionary.
   John Watkins, third edition (1807).

Commons Journal (1714).

Dictionary of National Biography.

Encyclopedia Britannica, eleventh edition.

Gentleman's Magazine.

A New and General Biographical Dictionary.
   John Whiston (et alia) (1761).

Parliamentary Debates, 1713-1714.

Philosophical Transactions.
   Lowthorp Abridgement (1705).

Barnish. John Rev., Orthodoxy and Heterodoxy 1710-1730, Oxford B.L.,
Bodleian M.S. B.D. C.1.

ADDENDUM TO BIBLIOGRAPHY

PRIMARY SOURCES

Bibliotheque Angloise.   Amsterdam (1717-28).

Burnet, G.:   History of his own Times (1724), 2 vols.

Cambden, W.:   Britannia, Second Edition (1789).

Dugdale (Sir William):   Antiquities of Warwickshire (1656).

Howell, T.B.:   A Complete Collection of State Trials,
                Volume XV (1812).

Hutton's Mathematical and Philosophical Dictionary (1795).

Nichols, J.:   Literary Anecdotes of the Eighteenth Century,
               9 vols. (1812-1815).

SECONDARY SOURCES

Cornforth, J.: Country Life, November 10th 1966.
               Article on Lyndon Hall, Rutland.

Lees, L.W.D.: Chronicles of a Suffolk Parish Church (1949).

WORKS OF REFERENCE

Penny Cyclopaedia (of the Society for the diffusion of
                Useful Knowledge), Vol. XXVII (1843),
                published Charles Knight & Co., London.

Rees, A.:   The Cyclopaedia or Universal Dictionary of
            Arts, Sciences and Literature (1819).

Victoria County History of Rutland.

INDEX OF NAMES MENTIONED IN THE TEXT

THE DEVELOPMENT OF SCIENCE

An Arno Press Collection

Electro-Magnetism. 1981

Gravitation, Heat, and X-Rays. 1981

Laws of Gases. 1981

Theory of Solutions and Stereo-Chemistry. 1981

The Wave Theory of Light and Spectra. 1981

Ackerknecht, Erwin H. Rudolf Virchow. 1953 *and* Schwalbe, J., editor. Virchow-Bibliographie, 1843-1901. 1901

Airy, George Biddell. Gravitation. 1834

Anderson, David L. The Discovery of the Electron. 1964

Beer, John. The Emergence of the German Dye Industry. 1959

Brown, Theodore. The Mechanical Philosophy and the "Animal Oeconomy." 1981

Candolle, Alphonse de. Histoire des sciences et des savants depuis deux siecles. 1885

Cheyne, Charles H.H. An Elementary Treatise in the Planetary Theory. 1883

Cohen, I. Bernard, editor. Andrew N. Meldrum. 1981

Cohen, I. Bernard, editor. The Conservation of Energy and the Principle of Least Action. 1981

Cohen, I. Bernard, editor. The Leibniz-Clarke Correspondence. 1981

Cohen, I. Bernard, editor. Studies on William Harvey. 1981

Coleman, William, editor Carl Ernst Von Baer on the Study of Man and Nature. 1981

Coleman, William, editor. French Views of German Science. 1981

Coleman, William, editor. Physiological Programmatics of the Nineteenth Century. 1981

Domson, Charles. Nicolas Fatio de Duillier and the Prophets of England. 1981

Donahue, William H. The Dissolution of the Celestial Spheres. 1981

Farrell, Maureen. William Whiston. 1981

Gardner, Walter M., editor. **The British Coal-Tar Industry.** 1915

Godfray, Hugh. **An Elementary Treatise on the Lunar Theory.** 1871

Graetzer, Hans G. and David L. Anderson. **The Discovery of Nuclear Fission.** 1971

Grimaux, Édouard. **Lavoisier: 1743-1794.** 1888

Hall, Diana Long. **Why Do Animals Breathe?** 1981

Hall, Maria Boas. **The Mechanical Philosophy.** 1981

Hannequin, Arthur. **Essai critique sur l'hypothèse des atomes dans la science contemporaine.** 1899

Harvey-Gibson, Robert J. **Outlines of the History of Botany.** 1919.

Heidel, William Arthur. **Hippocratic Medicine.** 1941

Heilbron, John L. **Historical Studies in the Theory of Atomic Structure.** 1981

Helm, Georg. **Die energetik.** 1898

Herschel, J.F.W. **Essays from the Edinburgh and Quarterly Reviews.** 1857

Hiebert, Erwin N. **Historical Roots of the Principle of Conservation of Energy.** 1962

Hilts, Victor L. **Statist and Statistician.** 1981

Hirschfield, John Milton. **The Academie Royale des Sciences (1666-1683).** 1981

Home, Roderick Weir. **The Effluvial Theory of Electricity.** 1981

Kendall, Maurice G. and Alison Doig. **Bibliography of Statistical Literature.** Three volumes. 1962, 1965 and 1968

Maier, Clifford L. **The Role of Spectroscopy in the Acceptance of the Internally Structured Atom, 1860-1920.** 1981

Meyer, Kirstine. **Die entwickelung des temperaturbegriffs im laufe der zeiten.** 1913

Milne-Edwards, Henri. **Introduction à la zoologie générale.** 1853

Morgan, Augustus de. **An Essay on Probabilities.** 1838

Mouy, Paul. **Le développement de la physique cartésienne 1646-1712.** 1934

Olmsted, J.M.D. **Francois Magendie.** 1944

Partington, J.R. and D. McKie. **Historical Studies on the Phlogiston Theory.** 1937, 1938 and 1939

Petit, Gabriel and Maurice Leudet. **Les allemands et la science.** 1916

Priestley, Joseph. **History and Present State of Discoveries Relating to Vision, Light, and Colours.** 1772

Quetelet, M.A. **Letters Addressed to H.R.H. The Grand Duke of Saxe Coburg and Gotha, on the Theory of Probabilities, as Applied to the Moral and Political Sciences.** 1849

Roe, Shirley A., editor. **The Natural Philosophy of Albrecht von Haller.** 1981

Sayili, Aydin. **The Observatory in Islam.** 1960

Schofield, Christine Jones. **Tychonic and Semi-Tychonic World Systems.** 1981

Schweber, S.S., editor. **Aspects of the Life and Thought of Sir John Frederick Herschel.** 1981

Shirley., John W., editor. **A Source Book for the Study of Thomas Harriot.** 1981

Struve, Friedrich George Wilhelm. **Études d'astronomie stellaire.** 1847

Turner, Dorothy Mabel. **History of Science Teaching in England.** 1927

Woolf, Harry. **The Transits of Venus.** 1959

Wurtz, Adolf. **A History of Chemical Theory, from the Age of Lavoisier to the Present Time.** 1869

Youmans, Edward L., editor. **The Correlation and Conservation of Forces.** 1865

Zloczower, A. **Career Opportunities and the Growth of Scientific Discovery in Nineteenth Century Germany.** 1981